Rhode Island Statewide Emergency Medical Services Protocols

Version 2018.03

These protocols are established by the Rhode Island Ambulance Service Coordinating Advisory Board.

These protocols and standing orders shall supersede all protocols and standing orders previously published.

Contains all protocols effective July 1, 2018.

_____

John Potvin

Chairperson, Ambulance Service Coordinating Advisory Board

_____

Kenneth A. Williams, MD, FACEP

Medical Director

Acknowledgements

Rhode Island Ambulance Service Coordinating Advisory Board
John Potvin, NRP, Chairperson, EMS Director, East Providence Fire Department, RISAFF
Michael DeMello, NRP, Vice Chairperson, Chief, Bristol Fire Department
Raymond Medeiros, AEMT-C, Secretary, Battalion Chief, North Providence Fire Department, RISAFF
Kenneth Williams, MD, Rhode Island Medical Society
Joseph Lauro, MD, American College of Emergency Physicians
Lynne Palmisciano, MD, American Academy of Pediatrics
Michael Connolly, MD, American College of Surgeons, Committee on Trauma
Scott Partington, Chief, Narragansett Fire Department, RI Association of Fire Chiefs
James Richard, NRP, Captain, Cumberland Rescue Service, RISAFF
Jason Umbenhauer, AEMT-C, Deputy Chief, Warwick Fire Department
Randall Watt, AEMT-C, Captain, Little Compton Fire Department
John Pliakas, MSN, NP, NRP, Emergency Nurses Association
Dawn Lewis, PhD, RN, EMT, Hospital Association of Rhode Island
Adam Reis, RN, EMT, Chief, Access Ambulance
Andrew Pappas, NRP, CEO, Southcoast Emergency Medical Services, Inc.
Bethany Gingerella, RN, NRP, Charlestown Ambulance-Rescue Service
Paul Casey, AEMT-C, Deputy Chief, Cranston Fire Department, RI Speaker of the House appointee
Derek Silva, AEMT-C, Lieutenant, Providence Fire Department, RI Senate President appointee
Gillian Cardarelli, NRP, Lieutenant, Providence Fire Department
Kimberly Perreault, AEMT-C, Captain, Potterville Fire Department
Christine Moniz, AEMT-C, North Scituate Fire Department, RI State Firemen's League
Kathleen Barton, Public Member
Tina Goncalves, Esq., Chief, Pawtucket Police Department, Public Member
Carolina Roberts-Santana, MD, MHA, Program Director, EMS for Children, RI Department of Health
Jason Rhodes, MPA, AEMT-C, Chief, Center for Emergency Medical Services, RI Department of Health

Special thanks to:
**Division of Preparedness, Response, Infectious Disease & Emergency Medical Services,
Center for Emergency Medical Services**
Utpala Bandy, MD, MPH, Medical & Division Director, RI State Epidemiologist
Christine Goulette, MAT, Chief Administrative Officer
Jason M. Rhodes, MPA, AEMT-C, Chief, Center for Emergency Medical Services
Kenneth Williams, MD, EMS Medical Director
Carolina Roberts-Santana, MD, MHA, Program Director, EMS for Children
Todd Manni, Program Planner, Center for Emergency Medical Services
Eric Rossmeisl, AEMT-C, Field Technician, Center for Emergency Medical Services
Elizabeth Vieira, EMS Licensing Aide, Center for Emergency Medical Services

John Pliakas, MSN, NP, NRP, Primary author

These individuals selflessly and enthusiastically provided their expertise and extensive effort to ensure this document is of the highest quality and adheres to contemporary medical standards. The State of Rhode Island is truly grateful for their dedication to the emergency medical services community and to the residents of our state.

Preface

Use of Protocols

These protocols delineate the scope of care for the three EMS provider levels recognized and licensed by the Rhode Island Department of Health. Level specific care is denoted by letters corresponding to each licensure level as below:

| E | **Denotes the scope of care for the Emergency Medical Technician (EMT) provider.** |

| A/C | **Denotes the scope of care for the Advanced EMT Cardiac provider.** |

| P | **Denotes the scope of care for the Paramedic provider.** |

| A C P | **Denotes the scope of care for Advanced EMT Cardiac and Paramedic Providers.** |

Care begins at level **E** for all providers. The scope of care for levels above the EMT provider is delineated in level respective sections as above (they do not build on each other). Example, the Paramedic level of care begins with the care delineated in the **E** section of a protocol and continues with the care delineated in the **P** section. If there are no subsequent level delineations beyond level **E** in a specific protocol, the care outlined in level **E** applies to all EMS provider levels.

If the directive "Consider" is used for an intervention (procedure, medication etc.), it is expected the provider will use his/her best judgement in considering its appropriateness as it relates to the patient at hand. In the event an intervention is deferred or executed, the provider(s) must have a reasonable and well thought out rationale for its deferment or execution.

The *Routine Patient Care Protocol* applies to all patients. More than one protocol may apply to a single patient. Providers are expected to utilize all applicable protocols when managing a patient. If a patient's presentation does not clearly meet the recognition criteria of a specific protocol, providers should consult MEDICAL CONTROL for guidance.

The **PEARLS** section contains useful information and in some cases, direction of care. Direction of care contained in the PEARLS section is considered part of the protocol and it is expected that providers will follow this direction as applicable.

Some protocols are adult or pediatric specific. These protocols are identified by the letter A following the protocol number for an adult specific protocol or by the letter P and a teddy bear logo for a pediatric specific protocol.

This document is embedded with hyperlinks identified by Blue underlined - text.

- The table of contents is blue but not underlined, yet contains hyperlinks to each individual protocol.
- Table of contents sections are hyperlinked to each individual section
- At the bottom of each protocol, clicking the table of contents sign returns to the section of the protocol, and the section title returns to table of contents.

These protocols are a "living document" maintained electronically by the Center for Emergency Medical Services (CEMS) and as such may be edited or updated as required at any time. A detailed review and editing of these protocols will occur on an annual basis and an updated version will be released every January. Suggestions and supporting evidence/ literature must be received by CEMS by March 31st of the preceding year for consideration for inclusion in the January revision. The intention is to include references, recommended reading and links to related educational materials in future editions.

E

1. Respond to the scene in a safe manner:
   - Use information available from the dispatcher, consider scene safety and pre-arrival assessment and treatment of the patient.
   - Request appropriate additional resources, including advanced life support (ALS) response / intercept if available.
   - Use of emergency warning devices and siren should be used with discretion and only as appropriate for the nature of the response and given information.

2. Approach the scene cautiously and assess scene safety:
   - If a hazard is identified, request appropriate assistance and maintain scene safety through appropriate measures including the use of personal protective equipment (PPE) as indicated. If the scene is unsafe, stage until hazards / threats are mitigated.
   - Utilize standard precautions for all patient contacts. Standard precautions include, depending on the anticipated degree or exposure, the use of gloves, gowns, mask, and eye protection or a face shield. Hand hygiene should be performed before and after each patient contact. Institute transmission based precautions as indicated in Table 1 below.
   - <u>Bring all necessary equipment to the patient</u>.

3. Utilize the *Multiple Patient Incident (Mass Casualty Incident) Protocol* if indicated based the number of patients and/or need for resources.

4. Determine if the patient(s) meet pediatric or adult criteria by age. A **pediatric patient** is defined as a patient less than 16 years of age. For patients ≤ 36 kg/80 lbs., utilize a pediatric dosing device (Broselow, Handtevy or other). If the Broselow Pediatric Emergency Resuscitation Tape is utilized, patients measuring beyond the length of the tape should receive weight based medication dosing until the age of 16 or their total body weight is ≥ 50kg. Some special needs patients may require continued use of pediatric protocols regardless of age.

5. For trauma patients, evaluate mechanism of injury (MOI) and employ spinal movement restriction precautions (SMRPs) if indicated following the *Spinal Movement Restriction Precautions Protocol.*

6. Obtain and document the patient's chief complaint (CC), history of the present illness (HPI), past medical history (PMH), current medications (including over-the-counter medications) and allergies to medications.

7. Perform a primary assessment and obtain vital signs. Vital signs at a minimum shall include blood pressure (BP); palpated pulse; respiratory rate; and oxygen saturation as measured by oximeter (SpO_2). Temperature (oral, rectal, axillary, or esophageal probe) should be obtained and documented when available and in all critically ill or injured pediatric patients.

8. Perform a secondary assessment (the secondary assessment may consist of a focused examination for isolated injuries).

9. Treat life-threatening conditions in the order in which they are identified. Manage as indicated per age appropriate protocol(s).

10. All patients shall have their level of pain assessed utilizing an age appropriate pain scale (see age appropriate *Patient Comfort Protocol*). This assessment shall be documented on the Patient Care Report.

11. Pain, nausea and vomiting should be managed following age appropriate *Patient Comfort Protocols.*

12. Provide airway management when indicated following age appropriate *Airway Management Protocols*. All intubated patients, including those undergoing interfacility transfer, must have the airway monitored with quantitative waveform capnography.

13. For patients with dyspnea/shortness of breath, chest pain/discomfort presumed to be of cardiac etiology and/or with a SpO_2 of <94%, oxygen should be administered via appropriate delivery device to maintain the SpO_2 in a range of 94-99%. During cardiopulmonary resuscitation (CPR) oxygen should be administered at the highest FiO_2 possible. Once return of spontaneous circulation (ROSC) is achieved, the FiO_2 should be titrated to the lowest concentration required to maintain the SpO_2 in a range of 94-99%.

14. Advanced life support (ALS) providers may establish intravenous (IV) access in any unstable or potentially unstable patient or when required for protocol directed therapeutic intervention (e.g. analgesia, treatment of nausea or vomiting etc.). 0.9% saline solution is the fluid of choice for "keep vein open" (KVO) purposes and limited fluid boluses. Unless otherwise directed by protocol, Lactated Ringers is the fluid of choice for significant fluid resuscitation for shock and burn injury. The KVO rate for the adult patient is 20-30 ml/hour and in the pediatric patient 10-20 ml/hr. Flow rates should be adjusted as directed by protocol. When appropriate, an intermittent needle therapy (INT) device (a.k.a. saline lock) may be placed in lieu of an infusing IV. Intraosseous (IO) access is interchangeable with IV access and reference to the IV route of medication administration is inclusive of the IO route.

15. Pediatric maintenance IV fluids should be calculated as follows: 4ml/kg for the first 10 kg + 2 ml per the second 10 kg plus 1 ml for each additional kg.

16. All intravenous infusions, unless exempted in Appendix 09.01, must be administered via an electronic infusion pump. See Appendix 09.01 for *Standard Concentrations for IV Admixtures*.

17. A multi-lead ECG (≥12 lead) should be obtained, if available, in any patient presenting with signs or symptoms suggestive of cardiac ischemia/infarction (chest pain/discomfort, known or suspected anginal equivalent, non-specific GI distress in an elderly female or diabetic, etc.) and in all patients post medical cardiac arrest with ROSC.

18. All procedures (successful and unsuccessful attempts) shall be documented on the Patient Care Report (PCR). Documentation shall include at a minimum: the procedure; device used; anatomic location; number of attempts; provider performing the procedure; and any noted complications.

19. Communicate with **MEDICAL CONTROL** as indicated and/or provide entry notification to receiving hospital facility. EMS providers may consult directly with a **MEDICAL CONTROL** physician at any time they feel such communication may be helpful in the care of a patient. Entry notification to the receiving hospital facility shall occur via electronic communication utilizing the RI Patient Tracking System (PTS). It may also occur via direct voice contact utilizing telephonic or radio communication.

20. Transport the patient to the nearest appropriate Hospital Emergency Facility (see Table 2 - Point of Entry - Specialized Hospital Emergency Facilities). Patients in respiratory or cardiac arrest should be transported to the nearest Hospital Emergency Facility unless otherwise directed by MEDICAL CONTROL. Providers should recognize and consider that, absent an acute life threatening condition or a condition requiring transportation to a specialized care facility, the most appropriate transport destination may be a facility at which the patient is followed or receives care (this facility may not be the geographically nearest).

21. Some patients who are beyond the pediatric age limit with chronic medical conditions may request or require transportation to a Pediatric Specialty Care Facility.

22. During transportation:
    - Stretcher patients shall be secured in an appropriate restraint system providing both transverse and longitudinal protection. Straps are required at the patient's knees, hips, and over the shoulders.
    - Ambulance cots shall be positioned at the lowest possible position during transportation and movement of the patient.
    - Seated patients shall be restrained with a lap and chest safety belt restraining system.
    - Pediatric patients of appropriate age, height or weight shall be transported utilizing a restraint system (child safety seat) compliant with Federal Motor Vehicle Safety Standards (FMVSS). The car safety seat shall be properly affixed to a stretcher with the head section elevated or to a vehicle seat, unless the patient requires immobilization of the spinal column, pelvis, or lower extremities; or the patient requires resuscitation or management of a critical problem.
    - All heavy items and equipment (e.g. cardiac monitors, oxygen cylinders) in the patient care compartment of the ambulance shall be secured following the manufactures specifications.
    - <u>Use of emergency warning devices and siren are to be restricted to use only while transporting patients with acute life threatening conditions requiring time sensitive interventions.</u>

23. The following may be used as a reference for pediatric age appropriate systolic blood pressure:

Neonate	>60 mmHg
Infant to 1 year	>70 mmHg
1-10 years	>70 + (2 x age in years)
>10 years	>90 mmHg

24. A face to face verbal report (hand-off) detailing the patient's chief complaint, abnormal assessment findings, EMS interventions, and the patient's response to treatment shall be given to a licensed health care provider at the receiving facility.

25. All care shall be documented in accordance with the *EMS Documentation Protocol* on a PCR.

26. Direct patient care provided by a non-transporting R.I. licensed ambulance (i.e. fire suppression unit, ALS intercept etc.) must be documented by the completion of a PCR by the licensed EMS providers staffing the unit and providing care. This may result in the generation of two PCRs within one service for the same patient/incident.

E

Table 1 - Transmission Based Precautions		
Category	**Category Components**	**Clinical Applicability**
Contact	Standard precautions + use of gloves during patient contact and when touching the patient's immediate environment or belongings, changing of gowns and gloves after every patient contact, cleaning and disinfecting all surfaces and reusable equipment, and strict hand washing.	Norovirus, rotavirus, enterovirus, *Clostridium difficile*, acute diarrhea of unknown etiology, respiratory syncytial virus (RSV), adenovirus, multidrug resistant organisms (MRSA, VRE, ESBLs, VISA/VRSA), scabies/head lice, herpes zoster (shingles) with disseminated disease, varicella (chickenpox), herpes simplex virus (HSV) in neonates, drainage (abscess, cellulitis, ulcer, burn or wound) not contained adequately by dressing or not dressed, impetigo or rash of unknown etiology.
Droplet	Standard precautions + standard surgical mask for providers accompanying patient in the back of the ambulance or within 3 feet of the patient and a surgical mask or non-rebreather mask on the patient.	Influenza, parainfluenza, parvovirus, meta-pneumovirus, rhinovirus, adenovirus, respiratory syncytial virus (RSV), bronchiolitis, meningitis, mumps, group A streptococcal pharyngitis in infants and children, infectious rash, pertussis (whooping cough) upper respiratory infections; including unknown infections prior to organism ID.
Airborne	Standard precautions + N95 respirator mask or powered air purifying respirator [PAPR]).	Tuberculosis (TB) or symptoms consistent with pulmonary TB (night sweats, fever, hemoptysis, unexplained weight loss), Mycobacterium tuberculosis, measles-rubeola, varicella (chickenpox), herpes zoster (shingles) with disseminated disease, vesicular (fluid filled) rash.
Notes		

- Hand hygiene is required before and after each patient contact for all patients.
- *Clostridium difficile* and acute diarrhea of unknown etiology require soap and water hand hygiene.
- Some pathogens require the use of a combination of precaution categories (i.e. varicella zoster (chickenpox) requires standard precautions + airborne precautions + contact precautions.

E

Table 2- Point of Entry - Specialized Hospital Emergency Facilities	
ACS Verified Level 1 Trauma Centers - Adult	Rhode Island Hospital
ACS Verified Level 1 Trauma Centers – Pediatric	Hasbro Children's Hospital
ABA Verified Adult Burn Centers	Rhode Island Hospital
ABA Verified Pediatric Burn Centers	Hasbro Children's Hospital
Comprehensive Stroke Centers	Rhode Island Hospital
Primary Stroke Centers	Kent Hospital Landmark Medical Center Miriam Hospital Newport Hospital Our Lady of Fatima Hospital Roger Williams Medical Center South County Hospital Day Kimball Hospital Lawrence and Memorial Hospital
Primary PCI Capable Hospitals	Kent Hospital Landmark Medical Center Miriam Hospital Rhode Island Hospital Charlton Memorial Hospital Lawrence and Memorial Hospital
Emergency Hyperbaric Oxygen Chamber	Kent Hospital Massachusetts General Hospital
Pediatric Specialty Care Hospital	Hasbro Children's Hospital
Mental Health Preferred Facilities	Landmark Medical Center Newport Hospital Our Lady of Fatima Hospital Rhode Island Hospital Roger Williams Medical Center
Opioid Use Disorder Preferred Facilities Level 1 or 2 as designated by RIDOH and BHDDH	Kent Hospital Miriam Hospital Newport Hospital Our Lady of Fatima Hospital Rhode Island Hospital Roger Williams Medical Center
Recovery Navigation Program or Mental Health and Opioid Use Disorder Facility	BH Link – 975 Waterman Avenue, East Providence, RI 401-414-5465 (LINK)

E

Recognition
- Documentation is the highest level of EMS professional accountability.
- The EMS patient care report (PCR) is part of the patient's permanent medical record and is often examined by other medical providers for important and valuable information.
- The detail and accuracy of a patient's PCR is reflective of the quality and credibility of the EMS provider(s) completing the documentation.
- EMS documentation establishes compliance with or deviation from the established standard of care.

1. All patient contacts with a R.I. licensed ambulance service must be documented.
2. All documentation must be performed electronically utilizing a National Emergency Medical Services Information System (NEMSIS) complaint software platform. The version will be determined by the Center for EMS.
3. The use of paper documentation is only acceptable in the event of electronic documentation failure. Paper documentation may not be used as a "placeholder" for electronic documentation.
4. Receiving facilities shall be provided a completed copy of the patient care report (PCR) proximate to patient arrival. In the event this is not possible, the PCR will be completed and provided to the receiving facility as soon as possible and no greater than 4 hours after patient handoff. An explanation in the narrative section of the PCR should explain any delay in the provision of a PCR to a receiving facility and should indicate the means used to deliver the report and confirm that it was received.
5. *PCR data shall be transmitted or otherwise provided to the Center of Emergency Medical Services (CEMS) in the timeframe specified in the Rules and Regulations Relating to Emergency Medical Services [R23-4.1EMS].*
6. In addition to required patient demographics, the following shall be documented in the narrative section of the PCR for each patient contact (if a particular data point is not applicable or is unobtainable, it should be documented as such):
   - Chief complaint;
   - Mechanism of injury (if applicable);
   - Source of information (if not from the patient);
   - History of present illness or injury:
   - Past medical history;
   - Family and social history (only if relevant);
   - Medications;
   - Medication or other significant allergies;
   - Physical examination findings (may be a focused examination for isolated injuries/complaints);
   - ECG interpretation and other point of care lab data (i.e. finger stick blood glucose);
   - Assessment (presumptive field diagnosis); and
   - All interventions and an evaluation of the effect or response.
7. All procedures and procedure attempts shall be documented. Documentation of the following is required for all procedures:
   - anatomic location (if applicable);
   - number of attempts;
   - +/- success;
   - name of the provider (s) performing or attempting the procedure;
   - complications;
   - and patient response (if applicable).
   - Documentation elements specified in individual procedure protocols shall also be included.

E

8.   Medication administration documentation shall include the name of the medication, dose administered, route, time of administration, name of the administering provider, medication effect, and any adverse effects.

9.   At least two complete sets of vital signs shall be documented for transported patients and one complete set for non-transported patients (pulse, respirations, auscultated blood pressure, and the SpO_2 at a minimum). Vital signs should be repeated and documented after drug administration, prior to patient transfer, and as needed during transport. For patients <3 y. o., blood pressure measurement is not required for all patients, but should be measured if possible, especially in critically ill patients in whom blood pressure measurement may guide treatment decisions. For critically ill or injured patients, vital signs should be documented at least every ten minutes and whenever significant changes are noted. The use of "WNL" is inappropriate and should not be utilized as a substitute for the documentation of measured vital signs.

10.   Any addendums to the PCR shall be dated, timed and signed by the author.

E

TABLE OF CONTENTS

1. The direction of prehospital care at the scene of an emergency is the responsibility of the most appropriately trained / senior highest licensed EMS provider present. Care at the scene should be provided in a collaborative nature amongst all EMS providers present in the best interest of the patient.

2. If a patient's **private physician** is present and assumes responsibility for the patient's care:
   - Defer to the orders of the private physician, providing the orders are within the scope of practice of the EMS provider(s).
   - If at any time the patient's private physician is not in attendance, EMS providers should provide care following the *Rhode Island Statewide EMS Treatment Protocols* and on-line medical direction.

3. If an **intervener physician** is present (physician who is not the patient's private physician) and has appropriately identified him/herself as a licensed physician and has demonstrated their willingness to assume responsibility for the patient's care and document their interventions:
   - Keep in mind that most intervener physicians are well intentioned.
   - Defer to the wishes of the physician.
   - If the treatment at the scene differs from that outlined in the *RI Statewide EMS Treatment Protocols* or there is disagreement between EMS providers and the intervener physician, on-line medical control should be established and the intervener physician should be put in communication with the medical control physician.
   - The on-line medical control physician has the option of managing the case entirely, working with the intervener physician, or allowing the intervener physician to assume responsibility.
   - If there is disagreement between the intervener physician and the on-line medical control physician, EMS providers shall defer to the on-line medical control physician.
   - If the treatment at the scene differs from that outlined in the *RI Statewide EMS Treatment Protocols* or there is disagreement between EMS providers and the intervener physician and on-line medical control cannot be established, the intervener physician should agree in advance accompany the patient to the hospital. However, in the event of a multiple patient incident (MCI), patient care needs may require the physician to remain at the scene.

E

Recognition:

- **Adult patient** without vital signs and with at least one of the following: rigor mortis (rigid stiffness of the body), fixed lividity, obvious injury incompatible with life (e.g. decapitation) or obvious changes of decomposition (i.e. bloating, skin slippage, extensive green or black skin discoloration).
- **Pediatric patient** without vital signs and with at least one of the following: obvious injury incompatible with life (e.g. decapitation) or obvious changes of decomposition (i.e. bloating, skin slippage, extensive green or black skin discoloration).

E

- Patients not meeting the above criteria must receive resuscitative care following the age appropriate *Cardiac Arrest Protocol* **unless** the patient or the patient's qualified health care decision maker pursuant to R23-4.11-MOLST has completed **Medical Orders for Life Sustaining Treatment (MOLST)** or the patient has **Comfort One status**. Manage patients with MOLST or Comfort One status per the *MOLST/Comfort One Protocol.*
- By recognizing the evidence of lifelessness (as above in recognition), EMS providers have made the determination of death. The determination of death by EMS providers does not constitute pronouncement or certification of death, which are the responsibility of a physician or licensed independent practitioner (nurse practitioner [advanced practice RN] or a physician assistant (PA) who at the time is practicing in a physician supervised role).
- Once a determination of death has been made, responsibility for the patient lies with local or state law enforcement. Law enforcement is responsible for contacting the Medical Examiner's Office. The body should not be removed from the scene and the scene should be disturbed as little as possible.
- EMS documentation must include the specific criteria on which the determination of death was made.
- Follow *Deceased Persons Protocol*.

PEARLS:

- Fixed lividity (purple/blue discoloration in gravitationally dependent parts of the body), does not change appearance with palpation.
- Cyanosis and skin changes associated with hypoperfusion should not be confused with fixed lividity.

Introduction

Emergency Medical Services providers have traditionally functioned as a mobile healthcare unit and are a logical resource for of providing healthcare to the community as an extension of the primary care network, provided a formal process is followed, as outlined in this protocol. This protocol enables an EMS service to form a coalition of partners for the purpose of providing community based health care services within the scope of practice of the EMS provider. A community that is experiencing a gap in healthcare services, as evidenced by a community needs assessment, may elect to utilize the capabilities of the EMS system in collaboration with primary care providers and/or a hospital. Only EMS providers who have met the requirements of this protocol may practice under these guidelines. <u>The Rhode Island Center for Emergency Medical Services of the Rhode Island Department of Health is available to provide consultation and support to Mobile Integrated Healthcare initiatives.</u>

Description of Mobile Integrated Healthcare (MIH)

- MIH is the provision of healthcare using patient-centered mobile resources in the out-of-hospital environment.
- The MIH concept is envisioned to be an organized system of services, based on local need.
- The services must be provided by EMT-Cardiac and Paramedic licensed providers who are integrated into the local health care system, working in a collaborative manner with physicians, mid-level practitioners, home care agencies and other community health team colleagues. Oversight all activities must be provided by emergency and primary care physicians.
- The purpose of the initiative is to address the unmet needs of individuals who are experiencing intermittent health care issues. It is not intended to address long-term medical or nursing case management, and is not a replacement for traditional home health services.

Requirements to establish a MIH in RI:

Each MIH entity must write and upload a plan to the documents section in the Rhode Island EMS Information System (RIEMSIS). The plan should include the following elements:

I. General Project Description

- ○ Describe the community/communities to be served, and name the MIH program (e.g. South County MIH)
- ○ The service's base location(s) to be employed,
- ○ The unmet community health needs that will be addressed
- ○ The current community health team members being partnered with, and
- ○ The methodology for addressing the need, including any enhancements of the EMS response system that will result.

II. Community Needs Assessment

- ○ The EMS agency, hospital, primary care providers and any other partners must provide a needs assessment that demonstrates the gap in healthcare coverage that the MIH program intends to fill.

III. Medical Direction

- ○ The MIH program must establish a collaborative working relationship between the EMS Medical Director or designee, who will be responsible for medical policy, operations and continuous quality improvement, and medical direction for MIH services.
- ○ The leadership role of the primary care provider in every patient-EMS-provider interaction plays a pivotal role in the design of MIH, and is a requirement.

A
C
P

IV. Patient Interaction Plan

- Describe the nature of anticipated patient- EMT-provider care interactions.
- Specify how the patient community will be educated to have realistic expectations of the EMT provider and these interactions.
- Define who will be providing the MIH services and how these services will fit with the regular EMS staffing of the Service.

VI. Training Plan

- Describe what training will be provided to enable the EMS providers to deliver the services described above.
- List the objectives and outcomes of the training plan.
- Document who will be responsible for training, oversight, and coordination as well as their qualifications. There must be a continuing education and credentialing process in place, with documentation of each EMS provider's participation.
- Such a process shall be approved by the EMS Service Medical Director(s).

VII. Quality Management Program and Data Collection

- The EMS Service shall conduct a quality management (QM) program specifically for the community healthcare program.
- The QM program will incorporate all the components of an EMS QM program as specified in the *Rules and Regulations Relating to Emergency Medical Services [R23-4.1EMS]*.
- Describe what data demonstrates the need for this project, if any.
- Describe the data to be collected to demonstrate the impact of this project on the population served.
- Describe the data reporting plan and how the RI Center for EMS will receive data and reports.

VIII. Documentation

- The EMS provider may at any time, using their own discretion, decide to activate the 911 system for emergency treatment and transport to appropriate care.
- Electronic patient care reports of all community healthcare patient encounters must be submitted to the requesting medical practice according to policies developed in coordination between the EMS Service, collaborating home health agency (if any) and medical practice.
- Copies of these records shall be maintained by the EMS Service, and be available for review by the RI Center for EMS.
- The EMS Service will participate in electronic data collection as required by the RI Center for EMS.

Scope and applicability

- The scope of Mobile Integrated Healthcare is to:
  - Provide urgent follow-up care when other home health resources and services are not an option.
  - Be part of a multidisciplinary team evaluating the needs of individuals who frequently utilize EMS in order to reduce that frequency while improving the quality and effectiveness of care so delivered;
  - Assess patients not recently seen by a physician, advanced practice registered nurse (APRN) or physician assistant (PA) for the need for chronic disease screening, and in collaboration with a primary care practice, help to initiate that screening.
  - By protocol and after real time consultation with a primary care practice professional, provide on scene treatment without transportation.

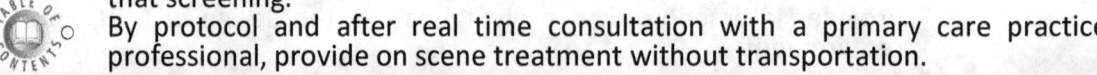

- Agencies wishing to provide MIH services must be a licensed emergency medical services agency pursuant to Rhode Island General Law §23-4.1 and the *Rules and Regulations Relating to Emergency Medical Services [R23-4.1EMS]*.
- Services shall have full commitment from the patient's primary care provider (PCP) to be involved in the MIH patient care process in the form of a memorandum of understanding (MOU) executed by both parties.
- The roles within the scope of MIH shall include the following, but not limited to:
  - Assessments within the scope of practice of the MIH practitioner, as ordered by the patient's PCP;
  - Screening and education for chronic disease in patients who are home bound when other resources are not available or have failed;
  - Injecting of immunizations and vaccinations for patients in the home (e.g. influenza vaccine) where the vaccine is provided by a primary care practice, transported appropriately, and only with a written order from the primary care physician for that patient.
  - Urgent laboratory specimen collection (i.e. blood draws and cultures), as requested by the primary care practice;
  - Public health functions, such as evaluation of patients enrolled in a direct active monitoring or direct observed therapy program prescribed by the Department of Health
  - By protocol and after real time consultation with primary care practice professional, provide on-scene treatment without transportation; and
  - Home safety assessments, including prevention of falls and fire safety.

Referral Process & Eligibility

- Patients eligible for MIH may be identified in the following ways:
  - Referral from the primary care provider (PCP).
  - Referral from emergency department (ED) physician upon discharge, provided that:
    ⇒ A memorandum of understanding is already in place with that emergency department and,
    ⇒ A primary care practice with an MOU is included in the follow-up plan at the time of the referral
  - Referral from a home nursing care provider or home care provider as a result of ineligibility for these services.
  - EMS initiated as a result of frequent EMS calls from specific patients with approval from the PCP.
  - RI Department of Health (RIDOH) referral for public health functions.
- Eligibility criteria for patients eligible for MIH include the following provisions:
  - Patient must be enrolled in the Rhode Island Health Information Exchange (Current Care or other health information exchange designated by the Director).
  - Patient must not be eligible for home nursing care or home care services unless referred by a PCP (e.g. pending initiation of home health services).
- Referrals and Identifying Potential MIH Patient

  - **PCP Referrals:**
    ⇒ The patient requires urgent assessment based on a phone assessment by the primary care practice;
    ⇒ The patient does not currently have access to or is ineligible for home nursing care or home care services, and requires frequent needs that can be addressed in the community via MIH.
    ⇒ The patient requires follow-up assessment / care that can be fulfilled through MIH.

A
C
P

A
C
P

Referral Process & Eligibility
- **EMS Initiated:** MIH patients must meet the eligibility criteria.
  - ⇒ To make a referral, the EMS Department shall contact the patient's PCP or case manager to suggest the referral.
  - ⇒ The patient does not currently have or is ineligible for home nursing care or home care services.
- **RIDOH Initiated:** by referral from various offices within the RI Department of Health.

Coordinating Care and MIH
- During each visit:
  - A call to the patient's home or residence should be made prior to a visit. If the patient is not home:
    - ⇒ The PCP must be notified;
    - ⇒ A visit should be rescheduled twelve (12) hours unless canceled by the PCP;
  - Services shall be provided by no less than one (1) licensed EMS practitioner educated in the concepts of MIH, as determined by the Department;
  - Prior to delivering care the MIH practitioner must verify the patient's identity;
  - The patient's PCP must be available for consultation during MIH visits; and
  - MIH practitioner has until the end of their shift to complete a patient visit report, which shall then be uploaded to Rhode Island EMS Information System and Health Information Exchange (Current Care or other health information exchange designated by the Director).

- **In the event that the MIH practitioner determines a medical emergency exists, or the visit becomes a medical emergency, the MIH agency shall provide emergency transport for the patient. The MIH practitioner shall notify the PCP en route if there is no time for prior notification.**

Field Treatment Without Transportation
- EMS practitioners, in direct consultation with the patient's PCP, are allowed to administer field treatment without transportation of the patient to a hospital facility.
- The PCP shall provide on-line medical control to the EMS practitioner to authorize field treatment, provided said treatment is within the scope of practice of the on-scene EMS practitioner and is consistent with the *Rhode Island Statewide EMS Protocols.*

Transportation to a non-hospital facility
- A patient may be transported to a non-hospital facility, including a health clinic, PCP office, urgent care facility or substance abuse treatment center only when:
  - Direct consultation with the patient's PCP provides authorization, or;
  - Certain clinical criteria have been met, particularly for transportation to a substance abuse treatment center, or;
  - Authorization is provided by medical control.

PEARLS:
- At no time are EMS practitioners allowed to perform activities not included in the skill level of their scope of practice. MIH functions expand the role of EMS practitioners, not the scope of practice.
- If at any time there is any doubt as to whether the patient needs care at a hospital-based emergency department or not, immediate transportation shall be provided to an appropriate hospital emergency facility.
- Communities participating in the MIH program are also strongly encouraged to be leaders and participants in improving community-based care, e.g., the HeartSafe Community program, opioid overdose prevention and blood pressure screenings, and partnering with primary care to coordinate MIH services.

Section 2: Medical Protocols

Recognition:
- Patient with altered mental status (GCS <8), abnormal motor posturing, unilateral or bilateral dilation of pupils, +/- bradycardia and/or hypertension.
- Possible etiologies include traumatic brain injury, epidural hematoma, subdural hematoma, subarachnoid hemorrhage, primary intracerebral hemorrhage, tumor, or encephalopathy.

E
- Routine patient care.
- If feasible, elevate the head of bed to > 30°.
- Avoid obstructions to venous drainage, such as tight cervical collars, securing devices for endotracheal tubes, or a non-midline head position.
- Provide airway management as indicated per the age appropriate *Airway Management Protocol.*
- Ventilate the patient to maintain an EtCO2 of 30-35 mmHg (avoid over aggressive hyperventilation), if capnography is not available, ventilate the patient at a rate of 14-16 bpm.
- Perform blood glucose analysis and manage hypoglycemia per the age appropriate *Diabetic Emergencies Protocol.*
- Transport the patient to the nearest appropriate Hospital Emergency Facility.

A
C
- Manage hypotension per the age appropriate *General Shock and Hypotension Protocol,* maintain the MAP ≥ 80 or the SBP ≥ 110.

P
- Manage hypotension per the age appropriate *General Shock and Hypotension Protocol,* maintain the MAP ≥ 80 or the SBP ≥ 110.
- 3% SALINE (HTS) 3 ml/kg (peds 1ml/kg) IV over 15 min.

PEARLS:
- Follow and document the patient's neurologic examination. Convey the patient's best neurologic examination and any episodes of hypoxia or hypotension during patient hand off.
- Attempt to obtain information related to use of antiplatelet or anticoagulants by the patient.
- Mild hyperventilation (EtCO2 30-35 mmHg) is a temporizing means of decreasing ICP and is reserved for patients with evidence of increased ICP/brain herniation. Brain ischemia is worsened by over aggressive hyperventilation. Patients without evidence of increased ICP should be ventilated to maintain normocapnia (EtCO2 35-45 mmHg).
- Short isolated episodes of hypoxia or hypotension should be avoided as they can cause secondary brain injury.
- Hyperglycemia is associated with worsened neurologic outcome. Glucose-containing solutions should be administered only as indicated for the treatment of hypoglycemia.

Recognition:
- Patient with complaint of abdominal pain, discomfort or cramping.

E
- Routine patient care.
- If the history and/or signs and symptoms are suggestive of a cardiac etiology, manage per the *Chest Pain- Acute Coronary-Syndrome -STEMI Protocol*.
- Manage hypotension, poor perfusion, or shock per the age appropriate *General Shock and Hypotension Protocol.*
- Transport the patient to the nearest appropriate Hospital Emergency Facility.

A C P
- Acquire a multi-lead ECG in any patient ≥ 35 yo. Manage per the *Chest Pain- Acute Coronary-Syndrome-STEMI Protocol* if findings suggest a cardiac etiology.
- Analgesia and antiemetic therapy as indicated per the age appropriate *Patient Comfort Protocol*.

PEARLS:

- Consider a possible cardiac etiology in patients ≥ 35 yo, diabetic patients and/or females especially with upper abdominal complaints or vague complaints of GI distress. Maintain a low threshold acquire a multi-lead ECG in these patients.
- Any female within child bearing age (12-50) should be managed as an ectopic pregnancy until such is ruled out.
- Abdominal aortic aneurysm should be considered in any patient ≥ 50 yo with abdominal pain, especially those with hypotension, poor perfusion, or shock.
- Mesenteric ischemia may present with severe pain with limited exam findings. Risk factors include age ≥ 60, atrial fibrillation, CHF and atherosclerosis.

Recognition:
- History of AI/Addison's disease, HIV/AIDS, sepsis.
- History of long term use of steroids (asthma, COPD, rheumatoid arthritis, organ transplant), use of antifungal agents.
- Hypotension, nausea, vomiting, dehydration, abdominal pain.

E
- Routine patient care.
- Maintain and promote normothermia by the use of blankets and increasing the ambient temperature if possible.
- Perform blood glucose (bG) analysis and treat as indicated per the age appropriate *Diabetic Emergencies Protocol.*
- Transport the patient to the nearest appropriate Hospital Emergency Facility.

A C P
- HYDROCORTISONE 100 mg (2 mg/kg, 100 mg max for peds) IV **or** METHYLPREDNISOLONE 125 mg (2 mg/kg, 60 mg max for peds) IV **or** DEXAMETHASONE 10 mg (0.3 mg/kg, 10 mg max for peds) IV.
- Manage hypotension/shock per the age appropriate *General Shock and Hypotension Protocol.*

PEARLS:

- Consider AI in patients with hypotension refractory to IV fluids and or vasopressors.
- Consider a administering a steroid stress dose to patients <u>with a history of AI</u> and any of the following: shock, fever (T >100.4) and ill appearing, multisystem trauma, burns (partial/full thickness) > 5% TBSA, environmental hypo or hyperthermia or vomiting or diarrhea with evidence of dehydration.

Recognition:
- History of exposure to an antigen (e.g. bee/wasp sting, shellfish, tree nuts, latex, medication)
- Itching, urticaria (hives), angioedema, wheezing, respiratory distress, chest or throat tightness, difficulty swallowing, GI symptoms, hypotension.

Symptom Severity Classification

Mild	Flushing, urticaria, itching, erythema with normal blood pressure and perfusion
Moderate	Flushing, urticaria, itching, erythema plus respiratory (wheezing, dyspnea, hypoxia) with normal blood pressure and perfusion
Severe	+/- skin symptoms depending on perfusion. Possible itching, erythema plus respiratory (wheezing, dyspnea, hypoxia) or gastrointestinal (nausea, vomiting, abdominal pain) with hypotension and poor perfusion

E
- Routine patient care.
- Assess symptom severity.
- For patients with symptoms of **mild severity**, monitor and reassess for worsening signs and symptoms.
- For patients with symptoms of **moderate severity**, consider EPINEPHRINE (1:1,000) 0.3 mg IM (lateral thigh) [auto-injector preferred] (avoid in patients > 50 yo or with a history of cardiac disease with mild symptoms only).
- For patients with **severe symptoms,** administer EPINEPHRINE (1:1,000) 0.3 mg IM (lateral thigh) [auto-injector preferred] (for patients > 50 yo or with a history of cardiac disease, administer 0.15 mg) every 5 min if no improvement to max of 3 doses. Additional doses require authorization from **MEDICAL CONTROL.**
- Transport the patient to the nearest appropriate Hospital Emergency Facility.

A
C
- For patients with symptoms of **mild severity**, DIPHENHYDRAMINE 50 mg PO/IV/IM.
- For patients with symptoms of **moderate severity**:
  - If indicated, continue IM administration of EPINEPHRINE (max 3 doses).
  - DIPHENHYDRAMINE 50 mg IV/IM if not already given PO.
  - ALBUTEROL 2.5-5 mg (+/- IPRATROPIUM) via SVN for continued wheezing (may repeat X 3).
  - METHYLPREDNISOLONE 125 mg IV **or** HYDROCORTISONE 100mg IV.
- For patients with **severe symptoms:**
  - If indicated, continue IM administration of EPINEPHRINE (max 3 doses).
  - NORMAL SALINE 500 ml IV bolus for a SBP <100 mmHg (may repeat to max of 2L).
  - DIPHENHYDRAMINE 50 mg IV/IM if not already given PO.
  - ALBUTEROL 2.5-5 mg (+/- IPRATROPIUM) via SVN for continued wheezing (may repeat X3).
  - METHYLPREDNISOLONE 125 mg IV **or** HYDROCORTISONE 100 mg IV.
  - Consider GLUCAGON 1-4 mg IV in patients taking a beta antagonist.

P

- For patients with symptoms of **mild severity,** DIPHENYDRAMINE 50 mg PO/IV/IM **and** FAMOTIDINE 20-40 mg PO/IV.
- For patients with symptoms of **moderate severity:**
  - If indicated, continue IM administration of EPINEPHRINE (max 3 doses).
  - DIPHENHYDRAMINE 50 mg IV/IM if not already given PO.
  - ALBUTEROL 2.5-5 mg (+/- IPRATROPIUM) via SVN for continued wheezing (may repeat X3).
  - FAMOTADINE 20-40 mg PO/IV.
  - METHYLPREDNISOLONE 125 mg IV **or** HYDOCORTISONE 100 mg IV **or** PREDNISONE 60 mg PO.
- For patients with **severe symptoms:**
  - If indicated, continue IM administration of EPINEPHRINE (max 3 doses).
  - NORMAL SALINE 500 ml IV bolus for a SBP <100 mmHg (may repeat to a max of 2L max).
  - DIPHENHYDRAMINE 50 mg IV/IM if not already given PO.
  - ALBUTEROL 2.5-5 mg (+/- IPRATROPIUM) via SVN for continued wheezing (may repeat X3).
  - FAMOTIDINE 20-40 mg PO/IV.
  - METHYLPREDNISOLONE 125 mg IV **or** HYDROCORTISONE 100 mg IV **or** PREDNISONE 60 mg PO.
  - Consider GLUCAGON 1-4 mg IV in patients taking a beta antagonist.
  - EPINEPHRINE 1-4 mcg/min by IV infusion for patients with hypotension refractory to IM EPINEPHRINE.
  - EPINEPHRINE (1:10,000) 0.1 mg [100 mcg] diluted in 10 ml 0.9% saline] slow IV for <u>peri-arrest</u> anaphylaxis (may repeat X2).

PEARLS:
- Recommended exam: mental status, skin, cardiac, pulmonary.
- If possible, patients with severe symptoms should remain in a supine position.
- Allergic reactions may occur with only respiratory and gastrointestinal symptoms without a rash or other skin symptoms/signs.
- **The use of an auto-injector is strongly recommended for the administration of IM epinephrine at all provider levels if available.**
- Patients with moderate or severe reactions should have IV access established and cardiac monitoring initiated.
- Patients > 50 yo, with a history of cardiac disease, or a heart rate > 150 are at risk for cardiac ischemia following the administration of epinephrine. These patients should have ongoing cardiac monitoring and a multi-lead ECG following the administration of epinephrine.
- Angioedema may be seen in patients taking ACE inhibitors (ACE-I) [lisinopril, ramipril, captopril, benzapril, quinapril, enalapril]. ACE-I induced angioedema results from an excessive accumulation of bradykinin. This is different then the histamine mediated angioedema associated with allergic/anaphylactic reactions. The use of antihistamines, corticosteroids and epinephrine offer no benefit in ACE-I related angioedema.
- ACE-I induced angioedema usually starts with focal swelling (e.g. isolated swelling of the tongue or lips). Patients with severe angioedema involving the tongue with airway compromise often require nasotracheal intubation.
- In the case of hereditary angioedema (HAE), like ACE-I related angioedema, the use of antihistamines, corticosteroids and epinephrine offer no benefit. Some patients with HAE are prescribed medication which may reverse it. Paramedics may assist the patient with or administer these medication per patient or packaging/prescription instructions.

Recognition:
- History of exposure to an antigen (e.g. bee/wasp sting, shellfish, tree nuts, latex, medication)
- Itching, urticaria (hives) angioedema, wheezing, respiratory distress, chest or throat tightness, difficulty swallowing, GI symptoms, hypotension.

Symptom Severity Classification

Mild	Flushing, urticaria, itching, erythemia with normal blood pressure and perfusion
Moderate	Flushing, urticaria, itching, erythemia plus respiratory (wheezing, dyspnea, hypoxia) with normal blood pressure and perfusion
Severe	+/- skin symptoms depending on perfusion. Possible itching, erythemia plus respiratory (wheezing, dyspnea, hypoxia) or gastrointestinal (nausea, vomiting, abdominal pain) with hypotension and poor perfusion

E

- Routine patient care.
- Assess symptom severity.
- For patients with symptoms of **mild severity**, monitor and reassess for worsening signs and symptoms.
- For patients with symptoms of **moderate severity**, consider EPINEPHRINE (1:1,000) 0.15 for patients 15-30 kg (33-66 lbs) or 0.3 mg for patients > 30 kg (66 lbs) IM [lateral thigh] (auto-injector preferred).
- For patients with **severe symptoms**, administer EPINEPHRINE (1:1,000) 0.15 for patients 15-30 kg (33-66 lbs) or 0.3 mg for patients > 30 kg (66 lbs) IM [lateral] thigh (auto-injector preferred) every 5 min if no improvement to max of 3 doses. Additional doses require authorization from **MEDICAL CONTROL.**
- Transport the patient to the nearest appropriate Hospital Emergency Facility.

A
C

- For patients with symptoms of **mild severity**, DIPHENHYDRAMINE 1 mg/kg PO/IV/IM (max 50 mg).
- For patients with symptoms of **moderate severity**:
  - If indicated, continue IM administration of EPINEPHRINE (max 3 doses).
  - DIPHENHYDRAMINE 1 mg/kg IV/IM if not already given PO (max 50 mg).
  - ALBUTEROL 2.5-5 mg (+/- IPRATROPIUM) via SVN for continued wheezing (may repeat X3).
  - METHYLPREDNISOLONE 2 mg/kg IV (max 60 mg) **or** HYDROCORTISONE 2 mg/kg IV (max 100 mg).
- For patients with **severe symptoms**:
  - If indicated, continue IM administration of EPINEPHRINE (max 3 doses).
  - NORMAL SALINE 20 ml/kg IV bolus, repeat as needed to achieve age appropriate BP (60 ml/kg max).
  - DIPHENHYDRAMINE 1 mg/kg IV/IM if not already given PO (max 50 mg).
  - ALBUTEROL 2.5-5 mg (+/- IPRATROPIUM) via SVN for continued wheezing (may repeat X3).
  - METHYLPREDNISOLONE 2 mg/kg IV (max 60 mg) **or** HYDROCORTISONE 2 mg/kg IV (max 100 mg).

P

- For patients with symptoms of **mild severity**, DIPHENHYDRAMINE 1 mg/kg PO/IV/IM (max 50 mg) **and** FAMOTIDINE 1 mg/kg IV (max 40 mg).
- For patients with symptoms of **moderate severity:**
  - If indicated, continue IM administration of EPINEPHRINE (max 3 doses).
  - DIPHENHYDRAMINE 1 mg/kg PO/IV/IM if not already given PO (max 50 mg).
  - ALBUTEROL 2.5-5 mg (+/- IPRATROPIUM) via SVN for continued wheezing. (may repeat X3).
  - METHYLPREDNISOLONE 2 mg/kg IV (max 60 mg) **or** HYDROCORTISONE 2 mg/kg IV (max 100 mg) **or** PREDNISONE/PREDNISOLONE (Orapred) 2 mg/kg PO (max 60 mg).
  - FAMOTIDINE 1 mg/kg IV (max 40 mg).
- For patients with **severe symptoms:**
  - If indicated, continue IM administration of EPINEPHRINE (max 3 doses).
  - NORMAL SALINE 20 ml/kg IV bolus, repeat as needed to achieve age appropriate BP (60 ml/kg max).
  - DIPHENHYDRAMINE 1 mg/kg IV/IM if not already given PO (max 50 mg).
  - ALBUTEROL 2.5-5 mg (+/- IPRATROPIUM) via nebulizer for continued wheezing (may repeat X3).
  - METHYLPREDNISOLONE 2 mg/kg IV (max 60 mg) **or** HYDROCORTISONE 2 mg/kg IV (max 100 mg) **or** PREDNISONE/PREDNISOLONE (Orapred) 2 mg/kg PO (max 60 mg).
  - FAMOTIDINE 1 mg/kg IV (max 40 mg).
  - EPINEPHRINE (1:10,000) 0.01 mg/kg (0.01 ml/kg) diluted in 10 ml 0.9% saline IV for peri-arrest anaphylaxis (may repeat X2).

PEARLS:
- For patients <15 kg (33 lbs) with moderate or severe symptoms, P level providers may consider/administer Epinephrine (1:1,000) 0.01 mg/kg IM.
- If possible, patients with severe symptoms should remain in a supine position.
- Allergic reactions may occur with only respiratory and gastrointestinal symptoms without a rash or other skin symptoms/signs.
- **The use of an auto-injector is strongly recommended for the administration of IM epinephrine at all provider levels if available.**
- Patients with moderate or severe reactions should have IV access established and cardiac monitoring initiated.
- Tachycardia (HR >150) is common following administration of epinephrine and/or albuterol. These patients should have ongoing cardiac monitoring and should have a multi-lead ECG acquired following the administration of epinephrine.
- Do not withhold epinephrine in a normal healthy child.
- Epinephrine is the single most effective drug to reverse immediate moderate and sever symptoms and should be the first drug administered.
- Diphenhydramine helps rash and itch, steroids take time for effects to be seen.
- Angioedema may be seen in patients taking ACE inhibitors (ACE-I) [lisinopril, ramipril, captopril, benzapril, quinapril, enalapril]. ACE-I induced angioedema results from an excessive accumulation of bradykinin. This is different then the histamine mediated angioedema associated with allergic/anaphylactic reactions. The use of antihistamines, corticosteroids and epinephrine offer no benefit in ACE-I related angioedema.
- ACE-I induced angioedema usually starts with focal swelling (e.g. isolated swelling of the tongue or lips). Patients with severe angioedema involving the tongue with airway compromise often require nasotracheal intubation.
- In the case of hereditary angioedema (HAE), like ACE-I related angioedema, the use of antihistamines, corticosteroids and epinephrine offer no benefit. Some patients with HAE are prescribed medication which may reverse it. Paramedics may assist the patient with or administer these medication per patient or packaging/prescription instructions.

Recognition:

- Patient with change in mental status from baseline

- Routine patient care.
- Perform blood glucose (bG) analysis. If the bG is ≤ 60 mg/dl or ≥ 250 mg/dl or signs and symptoms of hypoglycemia are present in the absence of the ability to perform bG analysis, treat patient per the *Diabetic Emergencies Protocol*.
- Obtain history and perform initial assessment to include mental status, neurologic, head, ears, eyes, nose, throat (HEENT), skin, lungs, cardiac, abdomen, back, extremities.
- Exit to appropriate protocol as indicated based on history and assessment findings:

Suggestive Findings	Protocol
Miosis, hypoventilation/apnea, needle track marks, other toxidrome findings	Toxicological Emergencies
Acetone odor on breath, rapid respiratory rate	Diabetic Emergencies
Hypotension or signs of poor perfusion	General Shock and Hypotension
Evidence of trauma, unequal pupils	Head Trauma - Traumatic Brain Injury
Hypothermia, hyperthermia	Hypothermia and Localized Cold Injury

E

Recognition:
- Patient with change in mental status from baseline

- Routine patient care.
- Perform blood glucose (bG) analysis. If the bG is ≤ 60 mg/dl or ≥ 250 mg/dl or signs and symptoms of hypoglycemia are present in the absence of the ability to perform bG analysis, treat patient per the _Diabetic Emergencies Protocol._
- Obtain history and perform initial assessment to include mental status, neurologic, head, ears, eyes, nose, throat (HEENT), skin, lungs, cardiac, abdomen, back, extremities.
- Exit to appropriate protocol as indicated based on history and assessment findings:

Suggestive Findings	Protocol
Miosis, hypoventilation/apnea, needle track marks, other toxidrome findings	Toxicological Emergencies
Acetone odor on breath, rapid respiratory rate	Diabetic Emergencies
Hypotension or signs of poor perfusion	General Shock and Hypotension
Evidence of trauma, unequal pupils	Head Trauma - Traumatic Brain Injury
Hypothermia, hyperthermia	Hypothermia and Localized Cold Injury

E

Recognition:
- An event occurring in an infant < 1 yo when the observer reports a sudden, brief (< 1 min), and now resolved episode of ≥ 1 of the following: (1) cyanosis or pallor (2) absent, decreased or irregular breathing (3) a marked change in tone (hyper or hypotonia) or (4) an altered level of responsiveness.

E

- Routine patient care.
- Perform Blood Glucose Analysis and manage per the *Diabetic Emergencies Protocol.*
- Obtain history of event with particular attention to:
  - Activity at onset and history of the event
  - State during the event (cyanosis, apnea, coughing, gagging, vomiting)
  - End of the event (duration, gradual or abrupt cessation, treatment provided)
  - State after the event (normal, not normal)
  - Recent history (illness, injuries, sick contacts, use of OTC medications, recent immunizations, new or different formula).
  - Past medical history (gestational age, pre-/perinatal history, GERD, seizures, previous BRUE).
  - Family history (sudden unexplained deaths, prolonged QT, arrhythmias).
  - Medications in the residence
  - Sleeping position/parent co-sleeping.
- Transport the patient to the nearest appropriate Hospital Emergency Facility.

PEARLS:
- BRUE was formerly known as Apparent Life Threatening Event (ALTE).
- BRUE is formally diagnosed (in the ED) only when there is no explanation for a qualifying event after conducting an appropriate history and physical examination.
- Recommended exam: general appearance, vital signs (including temperature), cardiac, pulmonary, skin, neurologic.
- BRUE is not a disease, but a symptom. Common etiologies include central apnea (immature respiratory center), obstructive apnea (structural), GERD (laryngospasm, choking, gagging), respiratory (pertussis, RSV), cardiac (CHD, arrhythmia), seizures.
- Always consider non-accidental trauma in any infant who presents with BRUE.
- Even with a normal physical examination at the time of EMS contact, patients that have experienced BRUE should be transported for further evaluation and work-up.
- It is important to note sleeping position as parent co-sleeping with child is associated with infant deaths.

Recognition
- This protocol applies to adult patients with pain or nausea or vomiting.
- This protocol is generally to be entered from a complaint specific protocol.

E
- Assess pain severity utilizing pain scale, circumstances, mechanism of injury, and severity of illness or injury.
- For **mild to moderate pain** (scale of 1-6), consider:
  - IBUPROFEN 10 mg/kg (typical adult 400-800 mg) PO **or**
  - ACETAMINOPHEN 15 mg/kg (typical adult 500-1000 mg) PO **or**
  - ASPIRIN 324-650 mg PO.
- For **moderate to severe pain** (scale >6), consider:
  - Interventions as above for mild pain.
  - If available, inhaled NITRONOX (50/50 nitrous oxide and oxygen blend).
- Monitor, reassess and document response to treatment.

A
C
- For **severe pain** (scale >6), FENTANYL 0.5-1 mcg/kg IV/IM/IN [max single dose 100 mcg] (may repeat every 10 min to a max of 300 mcg). IV doses should be given over 2 min.
- For patients with **nausea or vomiting**, ONDANSETRON 4 mg PO/IV/IM/ODT (may repeat X1 in 15 min).
- For patients requiring **electrical therapy** (cardioversion or pacing), consider MIDAZOLAM 2.5-5 mg IV/IM/IN **or** DIAZEPAM 2.5-5 mg IV/IM.
- Monitor and reassess response to treatment and vital signs prior to and 5 min following any dose of narcotic analgesic and before transfer of care (patient hand off). This must be documentation in the PCR.

P
- For **mild to moderate pain** (scale of 1-6), consider KETOROLAC 30 MG (15 mg if >65) IV **or** 60 mg (30 mg if >65) IM.
- For **severe pain** (scale >6):
  - FENTANYL 0.5-1 mcg/kg IV/IM/IN [max single dose 100 mcg] (may repeat every 10 min to a max of 300 mcg). IV doses should be given over 2 min **or**
  - For patients with traumatic pain or burns, KETAMINE 0.2 to 0.5 mg/kg IV **or** 0.5 to 1.0 mg/kg IM/IN (may repeat IV dosing X1 in 10 min and IM/IN dosing X1 30 min).
- For patients with **nausea or vomiting:**
  - ONDANSETRON 4 mg PO/IV/IM/ODT (may repeat X1 in 15 min).
  - For patients who do not respond to ONDANSETRON, consider PROMETHAZINE 6.25-12.5 mg IV/IM.
- For patients requiring **electrical therapy** (cardioversion or pacing), consider MIDAZOLAM 2.5-5 mg IV/IM/IN **or** DIAZEPAM 2.5-5 mg IV/IM **or** if IV access in unavailable in the patient requiring cardioversion, KETAMINE 2mg/Kg IM.
- For patients with an **advanced airway** in place (ETI/BIAD/cricothyrotomy) requiring sedation and analgesia, consider:
  - MIDAZOLAM 2-5 mg IV every 5-10 min as needed **or**
  - LORAZEPAM 1-2 mg IV may every 15 min as needed (max 10mg) **and**
  - FENTANYL 1-1.5 mcg/kg slow IV push.
- For patients with an **advanced airway** in place (ETI/BIAD/cricothyrotomy) if necessary for patient safety or to facilitate ventilation, consider ROCURONIUM 1 mg/kg IV **or** VECURONIUM 0.1 mg/kg IV **(must have continuous quantitative waveform capnography in place and must be preceded by sedation as above).**
- Monitor and reassess response to treatment and vital signs prior to and 5 min following any dose of narcotic analgesic and before transfer of care (patient hand off). This must be documented in the PCR.

PEARLS:

- DO NOT administer ibuprofen and ketorolac (Toradol) to patients that are pregnant, have a history of renal failure or transplant, are allergic to non-steroidal anti-inflammatory agents (NSAIDs), have active bleeding (including GI bleeding), have suspected intracranial hemorrhage, or in patients that may require surgical intervention such as those with open fractures/fractures with deformity.
- DO NOT administer aspirin to patients that have active bleeding (including GI bleeding, have suspected intracranial hemorrhage, or in patients that may require surgical intervention such as those with open fractures/ fractures with deformity).
- PO analgesics are not indicated for abdominal pain.
- DO NOT administer PO medications to patients that may require surgical intervention.
- Individual patients may respond differently to opioid analgesics. The patient's age, weight, clinical condition, co-administered/ingested drugs (alcohol, benzodiazepines) and prior exposure to opiates should all be considered when determining the dose to be administered. Weight based dosing provides a standard means for dose calculation, but does not predict patient response. Example: minimal doses of opioids may cause respiratory depression in elderly, opiate naïve or alcohol intoxicated patients.
- Avoid co-administering multiple sedating agents in non-intubated patients due to the risk for respiratory depression.
- Consider the use of waveform capnography in all patients receiving narcotic analgesics or ketamine.
- Patients with alcohol intoxication or those that have received benzodiazepines are at increased risk for respiratory depression following the administration of narcotic analgesics.
- Sub-anesthetic (low) dose ketamine has demonstrated significant analgesic efficacy without the adverse effects associated with higher doses. While uncommon, ketamine administration may result in laryngeal spasm and/or increased salivation. Laryngeal spasm is transient and can be managed with positive pressure ventilation if need be.
- As the dose related effect of ketamine transitions from analgesia to anesthesia, nystagmus emerges and as such, ketamine administration should be discontinued when nystagmus occurs.
- Ketamine should be administered over 60 seconds when given IV.
- Ketamine should not be used in patients with penetrating ocular injuries or known coronary artery disease.
- Droperidol has a sedating effect. Document mental status and vital signs prior to administration.
- Confirm IV patency with a saline flush prior to administering promethazine.
- **Advanced airway placement MUST be confirmed by the presence of a capnographic waveform (> 6 breaths) prior to the administration of rocuronium or vecuronium and <u>Continuous</u> airway monitoring with waveform capnography is required.**

Notes:

- For IM administration, the 100 mg/ml concentration of Ketamine is preferred.
- Ketamine in a concentration of 100 mg/ml must be diluted 1:1 with 0.9% saline, D5W or sterile water creating a 50 mg/ml concentration prior to IV use.

Recognition
- This protocol applies to pediatric patients with pain, nausea or vomiting.
- This protocol is generally to be entered from a complaint specific protocol.

E
- Assess pain severity utilizing age appropriate pain scale (numeric, Wong-Baker faces or FLACC scale), circumstances, mechanism of injury, and severity of illness or injury.
- For **mild to moderate pain** (scale of 1-6), consider:
  - IBUPROFEN 10 mg/kg (800 mg max) PO **or**
  - ACETAMINOPHEN 15 mg/kg (1000 mg max) PO.
- For **severe pain** (scale >6), consider:
  - Interventions as above for mild pain.
  - If available, inhaled NITRONOX (50/50 nitrous oxide and oxygen blend).
- Monitor, reassess and document response to treatment.

A C
- For **severe pain** (scale >6), FENTANYL 0.5-1 mcg/kg IV/IM/IN [max single dose 75 mcg] (may repeat every 10 min to a max cumulative dose of 150 mcg). IV doses should be given over 2 min.
- For patients with **nausea or vomiting,** ONDANSETRON 0.2 mg/kg (max dose 4 mg) PO/IV/IM/ODT (may repeat X1 in 15 min). Do not use in patients <3 mo of age.
- For patients requiring **electrical therapy** (cardioversion or pacing), consider MIDAZOLAM 0.1 mg/kg [2.5 mg max] IV/IM/IN **or** FENTANYL 2 mcg/kg [75 mcg max] IV/IM/IN.
- Monitor and reassess response to treatment and vital signs prior to and 5 min following any dose of narcotic analgesic and before transfer of care (patient hand off). This must be documented in the PCR.

P
- For **minor to moderate pain** (scale of 1-6), consider KETOROLAC 0.5 mg/kg IV/IM (max 30 mg).
- For **severe pain** (scale >6):
  - FENTANYL 0.5-1 mcg/kg IV/IM/IN [max single dose 75 mcg] (may repeat every 10 min to a max cumulative dose of 150 mcg). IV doses should be given over 2 min **or**
  - For patients with traumatic pain or burns, KETAMINE 0.1 mg/kg IV or 0.2 mg/kg IM/IN. IV dosing may be repeated every 10 min and IM/IN doses may be repeated once in 30 min as needed to a max cumulative dose of 10 mg.
- For patients with **nausea or vomiting**, ONDANSETRON 0.2 mg/kg (max dose 4 mg) PO/IV/IM/ODT (may repeat X1 in 15 min). Do not use in patients <3 mo of age.
- For patients requiring **electrical therapy** (cardioversion or pacing), consider MIDAZOLAM 0.1 mg/kg [2.5 mg max] IV/IM/IN **or** FENTANYL 2 mcg/kg [75 mcg max] IV/IM/IN or if IV access is unavailable, KETAMINE 2 mg/kg IM.
- For patients with an **advanced airway** in place (ETI/BIAD/cricothyrotomy) requiring sedation, consider FENTANYL 1.5-3mcg/kg IV.
- Monitor and reassess response to treatment and vital signs prior to and 5 min following any dose of narcotic analgesic and before transfer of care (patient hand off). This must be documented in the PCR.

PEARLS:

- Use extreme caution in administering opioids to patients < 10 kg.
- DO NOT administer Ibuprofen and ketorolac (Toradol) to patients that are pregnant, have a history of renal failure or transplant, are allergic to non-steroidal anti-inflammatory agents (NSAIDs), have active bleeding (including GI bleeding), have suspected intracranial hemorrhage, or in patients that may require surgical intervention such as those with open fractures/fractures with deformity.
- DO NOT administer aspirin to patients that have active bleeding (including GI bleeding, have suspected intracranial hemorrhage, or in patients that may require surgical intervention such as those with open fractures/ fractures with deformity).
- DO NOT administer PO medications to patients that may require surgical intervention.
- PO analgesics are not indicated for abdominal pain.
- Individual patients may respond differently to opioid analgesics. The patient's age, weight, clinical condition, co-administered/ingested drugs (alcohol, benzodiazepines) and prior exposure to opiates should all be considered when determining the dose to be administered. Weight based dosing provides a standard means for dose calculation, but does not predict patient response. Example: minimal doses of opioids may cause respiratory depression in elderly, opiate naïve or alcohol intoxicated patients.
- Avoid co-administering multiple sedating agents in non-intubated patients due to the risk for respiratory depression.
- Consider the use of waveform capnography in all patients receiving narcotic analgesics or ketamine.
- Patients with alcohol intoxication or those that have received benzodiazepines are at increased risk for respiratory depression following the administration of narcotic analgesics.
- **For pediatric administration, the 10 mg/ml concentration of ketamine is to be used.**
- Sub-anesthetic (low) dose ketamine has demonstrated significant analgesic efficacy without the adverse effects associated with higher doses. While uncommon, ketamine administration may result in laryngeal spasm and/or increased salivation. Laryngeal spasm is transient and can be managed with positive pressure ventilation if need be.
- As the dose related effect of ketamine transitions from analgesia to anesthesia, nystagmus emerges and as such, ketamine administration should be discontinued when nystagmus occurs.
- Ketamine should not be used in patients with penetrating ocular injuries or known coronary artery disease.
- Vomiting without diarrhea in pediatric patients may be related to pyloric stenosis, bowel obstruction or a CNS process (bleed, tumor, increased ICP).
- Utilize age appropriate pain scoring systems (see next page). For most patients > 9 yo, the numeric (1-10) scale is appropriate. For patients 2 mo-7 years, the FLACC scale may be used. he Wong-Baker-Faces scale may be used for patients > 3 yo.
- **Advanced airway placement MUST be confirmed by the presence of a capnographic waveform (> 6 breaths) prior to the administration of rocuronium or vecuronium. <u>Continuous</u> airway monitoring with side stream capnography is required.**

Wong-Baker FACES® Pain Rating Scale

0	2	4	6	8	10
No Pain	A Little Pain	A Little More Pain	Even More Pain	A Whole Lot Of Pain	Worst Pain

©1983 Wong-Baker FACES® Foundation. www.WongBakerFACES.org
Wording modified for adult use. Used with permission.

FLACC Scale

Categories	Scoring		
	0	1	2
Face	No particular expression of smile: disinterested	Occasional grimace or frown, withdrawn	Frequent to constant frown, clenched jaw, quivering chin
Legs	No position or relaxed	Uneasy, restless, tense	Kicking, or legs drawn up
Activity	Lying quietly, normal position, moves easily	Squirming, shifting back and forth, tense	Arched, rigid, or jerking
Cry	No crying (awake or asleep)	Moans or whimpers occasional complaint	Crying steadily, screams or sobs, frequent complaint
Consolability	Content, relaxed	Reassured by occasional touching, hugging, or talking to. Distractible	Difficult to console or comfort.
Each of the five categories (F) Face; (L) Legs; (A) Activity; (C) Cry; (C) Consolability is scored from 0-2, which enters in a total score between 0 and 10.			

Recognition:
- Shortness of breath, pursed lip breathing, wheezing/rhonchi, prolonged expiratory phase, use of accessory muscles, increased respiratory rate and effort, fever, cough.

E
- Routine patient care.
- For patients with **wheezing and history of asthma**, assist the patient with one dose of the patient's own rescue inhaler (use a spacer if available) **or** administer ALBUTEROL 2.5 mg (+/- IPRATROPIUM BROMIDE 500 mcg) via SVN. Contact **MEDICAL CONTROL** for authorization to administer additional doses.
- For patient not responding to initial therapy, consider continuous positive airway pressure (CPAP) at 5-10 cmH$_2$O.
- Transport the patient to the nearest appropriate Hospital Emergency Facility.

A C
- ALBUTEROL 2.5-5 mg (initial dose should include IPRATROPIUM 500 mcg, subsequent doses may be +/- IPRATROPIUM) via SVN (may repeat to a max of 4 doses).
- METHYLPREDNISOLONE 125 mg IV **or** HYDROCORTISONE 100 mg IV.
- For the patient with a diagnosis of **asthma who is in extremis**, contact **MEDICAL CONTROL** for authorization to administer EPINEPHRINE (1:1,000) 0.3 mg IM (lateral thigh) [auto-injector preferred] **or** TERBUTALINE 0.25 mg IM (may repeat every 15 min X2).

P
- ALBUTEROL 2.5-5 mg (initial dose should include IPRATROPIUM BROMIDE 500 mcg, subsequent doses may be +/- IPRATROPIUM) via SVN (may repeat PRN to a max of 4 doses).
- METHYLPREDNISOLONE 125 mg IV **or** HYDROCORTISONE 100 mg IV **or** PREDNISONE 60 mg PO **or** DEXAMETHASONE 10-20 mg IV/IM.
- Consider LEVALBUTEROL 1.25 mg via SVN or MDI (may repeat every 20 min as needed X4).
- For patients with asthma, consider MAGNESIUM SULFATE 2 gm IV over 10 min.
- For the patient with a diagnosis of asthma who is in extremis, EPINEPHRINE (1:1,000) 0.3 mg IM **or** TERBUTALINE 0.25 mg SC/IM (may repeat every 15 min X2).

PEARLS:
- Recommended exam: mental status, HEENT, skin, neck, cardiac, abdomen, pulmonary, extremities, neurologic.
- Differential diagnosis should include asthma, COPD, anaphylaxis, aspiration, pleural effusion, pneumonia, pulmonary embolus, CHF, hyperventilation, inhaled toxin, exit appropriate protocol based on index of suspicion.
- SpO$_2$ and EtCO$_2$ should be monitored continuously in patients with persistent distress.
- A "silent chest" in the asthmatic patient is a pre-respiratory arrest indicator.
- When considering endotracheal intubation in the asthmatic patient, if possible administer a rapid IV bolus of 1L NS or LR prior to intubation.
- Asthmatics are prone to hypotension following conversion to positive pressure ventilation (endotracheal intubation) due to decreased cardiac return (preload). Address volume status prior to intubation, following intubation ventilate the patient at a lower rate and utilize lower tidal volumes. Hypercapnia, as evidenced by increased EtCO2 levels is acceptable and attempts to correct with **hyperventilation should be avoided**.

Recognition:
- Respiratory distress with tachypnea, nasal flaring, retractions.
- Bronchiolitis: < 2 yo (peak 3-6 mo), +/- history of poor feeding/fussiness, increasing coryza and congestion, +/- low-grade fever, cough, fine crackles, fine/diffuse wheezing.
- Croup: 6 mo - 6 yo (peak <4 yo), upper respiratory tract infection, low grade fever, coryza, "barking" cough, varying degrees of respiratory distress.
- Asthma/RAD: > 2 yo, wheezing, cough, chest tightness, prolonged expiratory phase, +/- fever.
- Epiglottitis: > 2 yo, fever, toxic/ill appearing, favors upright position, drooling, stridor.

E

- Routine patient care.
- Allow patient to assume position of comfort.
- OXYGEN if the SpO_2 is ≤ 92%.
- For patients with **wheezing and history of asthma**, assist the patient with one dose of the patient's own rescue inhaler (use a spacer if available) **or** administer ALBUTEROL 2.5 mg (+/- IPRATROPIUM BROMIDE 500 mcg) via SVN. Contact **MEDICAL CONTROL** for authorization to administer additional doses.
- For patients with **suspected epiglottitis**:
  - Allow patient to assume positon of comfort.
  - Allow secretions to drain passively (i.e. avoid suctioning).
  - Minimize airway manipulation, provide airway management per age appropriate *Airway Management Protocols* only in the event of complete airway obstruction (most patients with epiglottis related airway obstruction can be effectively bag-mask ventilated).
- Transport the patient to the nearest appropriate Hospital Emergency Facility.

A C

- For patients **< 2 yo with continued respiratory distress and suspected bronchiolitis**:
  - Perform gentle nasopharyngeal suctioning for copious secretions.
  - If the patient appears to be dehydrated, consider NORMAL SALINE 20 ml/kg IV bolus.
- For patients **< 6 yo with continued respiratory distress and suspected croup**:
  - If the patient appears to be dehydrated, consider NORMAL SALINE 20 ml/kg IV bolus.
  - For patients with significant respiratory distress or stridor at rest, consider EPINEPHRINE (2.25% sol) 0.5 ml/3 ml NS via SVN (may repeat X1).
- For patients **≥ 2 yo with suspected reactive airway disease (RAD)/asthma**:
  - ALBUTEROL 2.5-5 mg (initial dose should include IPRATROPIUM 500 mcg, subsequent doses may be +/- IPRATROPIUM) via SVN for continued wheezing, (may repeat X3).
  - Consider IV access if the SpO_2 is ≤ 92% following the first dose of ALBUTEROL.
  - Consider continuous positive airway pressure (CPAP) @ 5-10 cmH20 if tolerated.
  - Consider METHYLPREDNISOLONE 2 mg/kg IV (max 60 mg) **or** HYDROCORTISONE 2 mg/kg IV (100 mg max).
  - For continued respiratory distress, contact **MEDICAL CONTROL** for authorization to administer EPINEPHRINE (1:1:000) 0.15-0.3 mg IM (lateral thigh) [auto-injector preferred] .

P

- For patients **< 2 yo with respiratory distress and suspected bronchiolitis**:
  - Perform gentle nasopharyngeal suctioning for copious secretions.
  - Supplemental oxygen if the SpO_2 is ≤ 92%.
  - If the patient appears to be dehydrated, consider a 20 ml/kg NORMAL SALINE IV bolus.

P

- Fore patients **< 6 yo with respiratory distress and suspected croup:**
  - Supplemental oxygen if the SpO_2 is ≤ 92%.
  - If the patient appears to be dehydrated, consider a 20 ml/kg NORMAL SALINE IV bolus.
  - DEXAMETHASONE 0.6 mg/kg PO/IM/IV [PO preferred] (max 10 mg).
  - For patients with significant respiratory distress or stridor at rest, consider EPINEPHRINE (2.25% sol) 0.5 ml/3 ml NS via SVN (may repeat X1).
- For patients **≥ 2 yo with reactive airway disease (RAD)/asthma:**
  - ALBUTEROL 2.5-5 mg (initial dose should include IPRATROPIUM 500 mcg, subsequent doses may be +/- IPRATROPIUM) via SVN for continued wheezing (may repeat X3).
  - Consider IV access if the SpO_2 is ≤ 92% following the first dose of inhaled beta agonist.
  - Consider continuous positive airway pressure (CPAP) @ 5-10 cmH_2O if tolerated.
  - METHYLPREDNISOLONE 2 mg/kg IV (max 60 mg) **or** HYDROCORTISONE 2 mg/kg IV (100 mg max) **or** PREDNISONE/PREDNISOLONE (Orapred) 2 mg/kg PO (60 mg max) **or** DEXAMETHASONE 0.6 mg/kg PO/IM/IV [PO preferred] (max 10 mg).
  - Consider MAGNESUM SULFATE 40 mg/kg IV over 10 min (2 gm max).
  - For continued respiratory distress, EPINEPHRINE (1:1:000) 0.15-0.3 mg IM (lateral thigh) [auto-injector preferred].

PEARLS:
- Differential diagnosis should include asthma/reactive airway disease, anaphylaxis, aspiration, foreign body airway obstruction, upper or lower airway infection (pneumonia), congenital heart disease, CHF, toxic ingestion.
- SpO_2 should be monitored continuously in patients with persistent distress.
- Bronchiolitis is a viral infection usually affecting infants and results in wheezing and typically does not respond to inhaled beta agonist. The management of bronchiolitis is largely supportive.
- Nebulized epinephrine is typically not recommended in bronchiolitis unless the patient is in extremis. Use of nebulized epinephrine prolongs EDs stays as it requires a 3 to 4 hour observation time after administration .
- Croup typically affects children from 6 mo to 6 yo with a peak at year < 4 yo and is characterized by a barking like cough resulting from upper airway edema. It is viral, may be associated with fever, is typically of gradual onset, and drooling is not typically noted.
- Use of nebulized epinephrine for croup should be limited to those patients with significant respiratory distress or stridor at rest.
- Epiglottitis is a bacterial infection typically affecting children > 2 yo. It is usually of rapid onset, associated with fever and drooling. Stridor may be present and the patient may assume the tripod position. Airway manipulation may worsen the condition and should be avoided. The incidence of epiglottis has decreased over the last couple of decades due to routine immunization against H. influenza.
- A "silent chest" in the asthmatic patient is a pre-respiratory arrest indicator.
- When considering endotracheal intubation in the <u>asthmatic</u> patient, if possible administer a rapid 20 ml/kg IV bolus of NS or LR prior to intubation.
- Asthmatics are prone to hypotension following conversion to positive pressure ventilation (endotracheal intubation) due to decreased cardiac return (preload). Address volume status prior to intubation, following intubation ventilate the patient at a lower rate and utilize lower tidal volumes. Hypercapnia, as evidenced by increased EtCO2 levels is acceptable and attempts to correct with **hyperventilation should be avoided**.

Recognition:
- Patient exhibiting any one or a combination of the following: anxiety, agitation, affect change, hallucinations, delusional thoughts, bizarre behavior, combative or violent behavior, expression of suicidal/homicidal thoughts.

E

- Consider safety of EMS providers first. If law enforcement is not present at the scene, consider staging in a safe area until the arrival of law enforcement.
- Routine patient care.
- Consider possible medical etiologies, managing per the following age appropriate protocols as indicated:
  - *Altered Mental Status*
  - *Diabetic Emergencies*
  - *Excited Delirium*
  - *Head Trauma - Traumatic Brain Injury*
  - *Toxicological Emergencies*
  - *Patient in Police Custody*
- Remove patient from stressful environment.
- Utilize the SAFER model:
  Stabilize the situation by lowering stimuli, including voice.
  Assess and acknowledge crisis by validating the patient's feelings and not minimizing them.
  Facilitate identification and activation of resources (clergy, family, friends, and police).
  Encourage patient to use resources and take action in the patient's best interest.
  Recovery/referral – transport patient to a Hospital Emergency Facility. If the patient is not transported, be sure the patient is in the care of a responsible individual or professional.
- Consider *Patient Restraint Procedure* if indicated (aggressive, agitated, psychosis, possible danger to self or others). Restraint should be performed/ assisted by law enforcement when available.
- Transport the patient to the nearest appropriate Hospital Emergency Facility.
  - If there are no medical or substance use disorder etiologies, transport patient to a mental health preferred facility if the transport time is less than 20 minutes (see Table 2 Point of Entry – Specialized Hospital Emergency Facilities in Routine Patient Care).
  - For patients with known or suspected opioid overdose, that are stable with adequate ventilation (before or after NALOXONE administration), transport patient to an opioid use disorder preferred facility, if the transport time is less than 20 minutes (see Table 2 Point of Entry – Specialized Hospital Emergency Facilities in Routine Patient Care).
- For patients with known or suspected alcohol or opioid use disorder, that are stable with adequate ventilation, consider transport to the recovery navigation program or mental health and opioid use disorder facility (see Table 2 Point of Entry – Specialized Hospital Emergency Facilities in Routine Patient Care), if the transport time is less than 20 minutes and by following the Alternative Transportation Algorithm.
- For patients with known mental health history and presenting with an acute exacerbation of their condition, without danger to self or others, consider transport to the recovery navigation program or mental health and opioid use disorder facility (see Table 2 Point of Entry – Specialized Hospital Emergency Facilities in Routine Patient Care), if the transport time is less than 20 minutes and by following the Alternative Transportation Algorithm.

A C

- For patients ≥16 with **aggressive or agitated behavior, without suspected acute substance abuse (alcohol)** who are not responsive to the above interventions, consider chemical restraint with MIDAZOLAM 2.5-5 mg IV or 5 mg IM/IN, may repeat PRN if the SBP >100 to a cumulative dose of 10 mg (5 mg if age ≥65).

P

- For patients ≥16 yo with **aggressive or agitated behavior** who are not responsive to the above interventions, consider chemical restraint with:
  - ○ HALOPERIDOL 5 mg IV/IM **or** DROPERIDOL 5 mg IV/IM (2.5 mg if age ≥ 65), may repeat either to a cumulative dose of 10 mg **or**
  - ○ MIDAZOLAM 2.5-5 mg IV **or** 5 mg IM/IN, may repeat PRN if SBP >100 to a cumulative dose of 10 mg (5 mg if age ≥ 65, avoid if suspected alcohol intoxication) **or**
  - ○ KETAMINE 4 mg/kg IM (max 400 mg) or 2 mg/kg IV (max 200 mg), for hypersalivation following KETAMINE administration, consider ATROPINE SULFATE 0.5 mg IV/IM.

PEARLS:

- Be sure to consider all possible medical/trauma etiologies for behavior (hypoglycemia, toxicological, hypoxic, head injury).
- Do not position or transport any restrained patient in such a way (e.g. prone) that could negatively affect the patient's respiratory or circulatory status.
- Any patient who is handcuffed or restrained by law enforcement and transported by EMS must be accompanied by a law enforcement officer.
- Continuous monitoring of $EtCO_2$ (waveform) and SpO_2 are mandatory in patients who receive physical or chemical restraint. When clinically feasible, ECG monitoring is required for patients that have received haloperidol.
- It may be necessary/appropriate to administer IM injections through clothing in extremely agitated patients.
- Extrapyramidal reactions associated with haloperidol or droperidol should be managed per the _Toxicological Emergencies Protocol_.
- **BH Link - 401-414-5465 (LINK)**

Alternative Transportation Algorithm

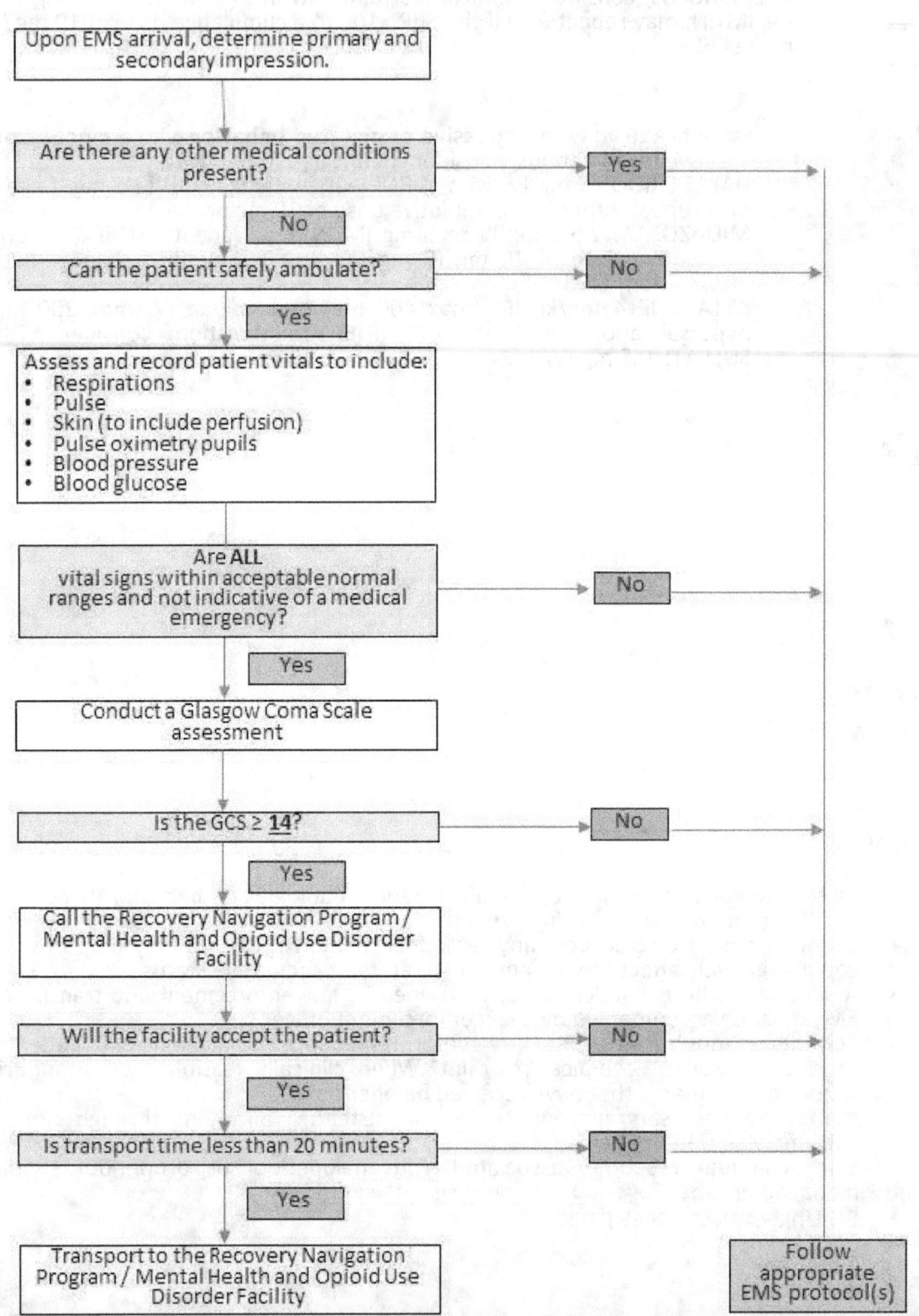

Upon EMS arrival, determine primary and secondary impression.

Are there any other medical conditions present? → Yes →

No ↓

Can the patient safely ambulate? → No →

Yes ↓

Assess and record patient vitals to include:
- Respirations
- Pulse
- Skin (to include perfusion)
- Pulse oximetry pupils
- Blood pressure
- Blood glucose

Are **ALL** vital signs within acceptable normal ranges and not indicative of a medical emergency? → No →

Yes ↓

Conduct a Glasgow Coma Scale assessment

Is the GCS ≥ <u>14</u>? → No →

Yes ↓

Call the Recovery Navigation Program / Mental Health and Opioid Use Disorder Facility

Will the facility accept the patient? → No →

Yes ↓

Is transport time less than 20 minutes? → No →

Yes ↓

Transport to the Recovery Navigation Program / Mental Health and Opioid Use Disorder Facility

Follow appropriate EMS protocol(s)

2.10 Adult   Diabetic Emergencies

Recognition:
- Hypoglycemia: anxiety, altered mental status, diaphoresis, seizures, tachycardia.
- Diabetic ketoacidosis: warm dry skin, tachycardia, rapid shallow breathing, hypotension/shock, acetone odor on breath, EtCO2 ≤29 mmHg.

- Routine patient care.
- Perform blood glucose (bG) analysis and determine bG level if not previously determined.
- Treat as below (in the absence of the ability to determine bG, patients with signs or symptoms of hypoglycemia should be treated as below as appropriate for their mental status and their ability to receive ORAL GLUCOSE SOLUTION).

bG	E	ACP
≤ 60 mg/dl **Patient is Awake and Alert**	1. ORAL GLUCOSE SOLUTION* (15 gm glucose) PO. 2. Recheck bG in 15 minutes. 3. Repeat dose of ORAL GLUCOSE SOLUTION if bG ≤ 60 mg/dl.	1. ORAL GLUCOSE SOLUTION* (15 gm glucose) PO. 2. Recheck bG in 15 minutes. 3. Repeat dose of ORAL GLUCOSE SOLUTION if bG ≤ 60 mg/dl.
≤ 60 mg/dl **Patient is Not Alert to Verbal Stimuli or is Nauseated/ Vomiting**	Contact **MEDICAL CONTROL** for authorization to administer GLUCAGON 1mg (1U) IM.	1. Thiamine 100 mg IV/IM. 2. D10W 250 ml (25g) IV over 5 minutes **or** DEXTROSE 50% (25 gm/50 ml) IV. Either may be repeated in 5 minutes if bG <60 mg/dl. 3. If unable to establish IV access, GLUCAGON 1mg (1U) IM.
60 -249 mg/dl	Recheck bG if condition changes.	Recheck bG if condition changes.
≥ 250 mg/dl	Recheck bG if condition changes.	1. If patient is dehydrated without evidence of CHF/fluid overload, administer NORMAL SALINE 500 ml IV bolus. Repeat as needed. 2.Infuse NORMAL SALINE at 150 ml/hr.

*Alternate glucose sources may be administered. Oral glucose equivalents include 3-4 glucose tablets, 4 oz. fruit juice (e.g. orange juice), non-diet soda, 1 tablespoon of pure RI maple syrup, sugar, or honey.

- Thiamine should be reserved for patients that have a history of alcohol abuse or who appear malnourished.
- If the patient is hypotensive, manage per age appropriate *General Shock and Hypotension Protocol.*
- Encourage patients who refuse transport after improvement in GCS and are back to baseline to consume complex carbohydrates (15 grams) and protein (12-15 grams) such as peanut butter toast, mixed nuts, milk or cheese to stabilize blood sugar.

- Transport the patient to the nearest appropriate Hospital Emergency Facility.

PEARLS:
- Patients who presented with a bG≤ 60 mg/dl and are taking oral hypoglycemic agents should be strongly encouraged to agree to transport.

Recognition:
- Hypoglycemia: anxiety, altered mental status, diaphoresis, seizures, tachycardia.
- Diabetic ketoacidosis: warm dry skin, tachycardia, rapid shallow breathing, hypotension/shock, acetone odor on breath, EtCO2 ≤29 mmHg.

- Routine patient care.
- Perform blood glucose (bG) analysis and determine bG level if not previously determined.
- Treat as below (in the absence of the ability to determine bG, patients with signs or symptoms of hypoglycemia should be treated as below as appropriate for their mental status and their ability to receive ORAL GLUCOSE SOLUTION).

bG	E	ACP
≤ 60 mg/dl **Patient is Awake and Alert**	1.ORAL GLUCOSE SOLUTION* (15 gm glucose) PO. 2. Recheck bG in 15 minutes. 3. Repeat dose of ORAL GLUCOSE SOLUTION if bG≤ 60 mg/dl.	1.ORAL GLUCOSE SOLUTION* (15 gm glucose) PO. 2. Recheck bG in 15 minutes. 3. Repeat dose of ORAL GLUCOSE SOLUTION if bG≤ 60 mg/dl.
≤ 60 mg/dl **Patient is Not Alert to Verbal Stimuli or is Nauseated/ Vomiting**	Contact **MEDICAL CONTROL** for authorization to administer GLUCAGON 0.1 mg/kg [1 mg max].	1. D10W 5 ml/kg over 5 minutes **or** DEXTROSE 25% (12.5 gm/50 ml) 2 ml/kg IV. Either may be repeated in 5 minutes if bG <60 mg/dl. 2. If unable to establish IV access, GLUCAGON 0.1 mg/kg IM [1 mg max].
60 -249 mg/dl	Recheck bG if condition changes.	Recheck bG if condition changes.
≥ 250 mg/dl	Recheck bG if condition changes.	1. If patient is dehydrated without evidence of CHF/fluid overload, administer NORMAL SALINE 20 ml/kg IV bolus. 2. Infuse NORMAL SALINE at 20 ml/hr.

*Alternate glucose sources may be administered. Oral glucose equivalents include 3-4 glucose tablets, 4 oz. fruit juice (e.g. orange juice), non-diet soda, 1 tablespoon of pure RI maple syrup, sugar, or honey.

- If the patient is hypotensive, manage per age appropriate *General Shock and Hypotension Protocol.*
- Encourage patients who refuse transport after improvement in GCS and are back to baseline to consume complex carbohydrates (15 grams) and protein (12-15 grams) such as peanut butter toast, mixed nuts, milk or cheese to stabilize blood sugar.
- Transport the patient to the nearest appropriate Hospital Emergency Facility.

Recognition:
- Volume overload: shortness of breath, dyspnea on exertion, crackles, JVD, peripheral edema.
- Hyperkalemia: muscle weakness, cardiac arrhythmias, peaked T waves, wide QRS complex/ sine wave ECG.

E
- Routine patient care.
- For patients with hemorrhage from a shunt/fistula, manage per the *External Hemorrhage Control Protocol.*
- For patients with pulmonary edema, manage per the *Acute Decompensated Heart Failure – Pulmonary Edema Protocol.*
- Transport the patient to the nearest appropriate Hospital Emergency Facility.

A C
- For patients that have had hemodialysis within the past 4 hours with a SBP < 90, and clear lung sounds, NORMAL SALINE 250ml IV BOLUS (repeat as needed if the lungs remain clear [max 1 L]).
- For patients in cardiac arrest or with suspected hyperkalemia, manage per the age appropriate *Cardiac Arrest Protocol* (if applicable) and contact **MEDICAL CONTROL** for authorization to administer CALCIUM CHLORIDE 1 gm IV **or** CALCIUM GLUCONATE 3 gm IV **and** SODIUM BICARBONATE 50 mEq IV.

P
- For patients that have had hemodialysis within the past 4 hours with a SBP < 90, and clear lung sounds, NORMAL SALINE 250ml IV BOLUS (repeat as needed if the lungs remain clear [max 1 L]).
- For patients in cardiac arrest or with suspected hyperkalemia, manage per the age appropriate *Cardiac Arrest Protocol* (if applicable) and consider CALCIUM CHLORIDE 1 gm IV **or** CALCIUM GLUCONATE 3 gm IV **and** SODIUM BICARBONATE 50 mEq IV.

PEARLS:
- Do not take a blood pressure or establish IV access in an extremity with a shunt/fistula in place.
- Access of a shunt or dialysis catheter in indicated only in the periarrest/cardiac arrest patient only when no other access is available. IO access should be attempted/utilized if available.
- Consider hyperkalemia in all dialysis or renal failure patients.
- Renal failure patients typically have multiple coexisting medical problems. Cardiac disease and hypertension are most prevalent in this population.
- Sodium bicarbonate and calcium preparations are not compatible and should be given through separate IV lines if possible. If they must be administered via the same IV line, the line should be flushed in between the administration of each.
- If locally applied pressure does not control hemorrhage from a dialysis fistula, a tourniquet should be utilized. The tourniquet should be applied as far away from the fistula as possible.

Recognition:
- New Onset (<24hrs)
  - Unilateral motor weakness or paralysis (including facial droop)
  - Unilateral numbness
  - Speech/language disturbance
  - Visual disturbance
  - Abrupt gait disturbance

E

- Routine patient care.
- Determine blood glucose level (bG) level, manage per the age appropriate *Diabetic Emergencies Protocol* if indicated.
- Determine the time of symptom onset (the last time the patient was seen normal). Every effort should be made to be precise in this determination. If the patient awoke with neurologic symptoms and was normal when they went to sleep, the last time seen normal should be documented as the time they went to sleep.
- Perform neurologic assessment and calculate the patient's Los Angeles Motor Scale (LAMS) [applies to new [<24 hr] neurologic deficits only]:

Facial Droop	
Absent	0
Present	1
Arm Drift	
Absent	0
Drifts Down	1
Fall Rapidly	2
Grip Strength	
Normal	0
Weak grip	1
No Grip	2

- **For patients with a LAMS score of ≥ 4 and the transport time is less than 30 minutes to a Comprehensive Stroke Center (CSC), transport the patient to the nearest CSC, otherwise transport the patient to the nearest stroke center.**
- Provide early notification to the stroke center and include the LAMS score for all suspected ischemic stroke patients. A LAMS score of ≥ 4 is suggestive of an Emergent Large Vessel Occlusion (ELVO). Patients with ELVO are best managed by a combination of thrombolysis and emergent embolectomy. These are time sensitive interventions. Embolectomy is only available at CSCs.
- On scene time should be limited to ≤ 10 minutes.
- Provide the receiving facility with an early **CODE STROKE** alert.
- If able, obtain the name of the patient's next of kin/healthcare proxy and their contact information (cell phone etc.). This information may be required to obtain consent for treatment at the receiving facility.
- Attempt to determine if the patient is taking oral anticoagulants [warfarin (Coumadin), apixaban (Eliquis), dabigatran (Pradaxa), rivaroxban (Xarelto), edoxaban (Savaysa)] and any recent history of surgery, trauma or seizures. This information should be documented on the Patient Care Report.
- It may be requested that the patient be transported directly to the radiology department at the receiving facility in an effort to expedite critical time sensitive imaging. EMS providers should comply with such requests.

P

- If established, IV access should be placed above the wrist (establishing IV access should not delay transport).
- Paramedics may transport patients receiving fibrinolytic agents (tPA), NICARDAPINE and LABETALOL via IV infusion. Infusions of NICARDIPINE and LEBETALOL may be titrated as per protocol 2.13 or following parameters set by the sending MD. These patients should also be manage per protocol 2.13.

PEARLS:
- Stroke requires time sensitive interventions. Each minute that passes during an acute stroke, approximately 2 million neurons are lost!
- Establishing IV access in the field or during transport may expedite the time to intervention at the receiving center, however this should not delay transport.
- The following conditions may mimic ischemic stroke: alcohol intoxication, cerebral infectious process, toxic ingestions, hypoglycemia, metabolic disorders, migraines, neuropathies (Bell's palsy), seizures, post-seizure persistent neurological deficits (Todd's paralysis), neoplasms, and hypertensive encephalopathy.
- Right hemispheric strokes may present with dysarthria (slurred speech), left sided hemiparesis /paralysis, left sided neglect, right sided gaze preference.
- Left hemispheric stroke may present with aphasia, impaired comprehension, right sided hemiparesis/paralysis, left gaze preference.
- Brainstem strokes may present with nausea, vomiting, vertigo, impaired speech, dysphasia, abnormal eye movements, and decreased level of consciousness, crossed findings.

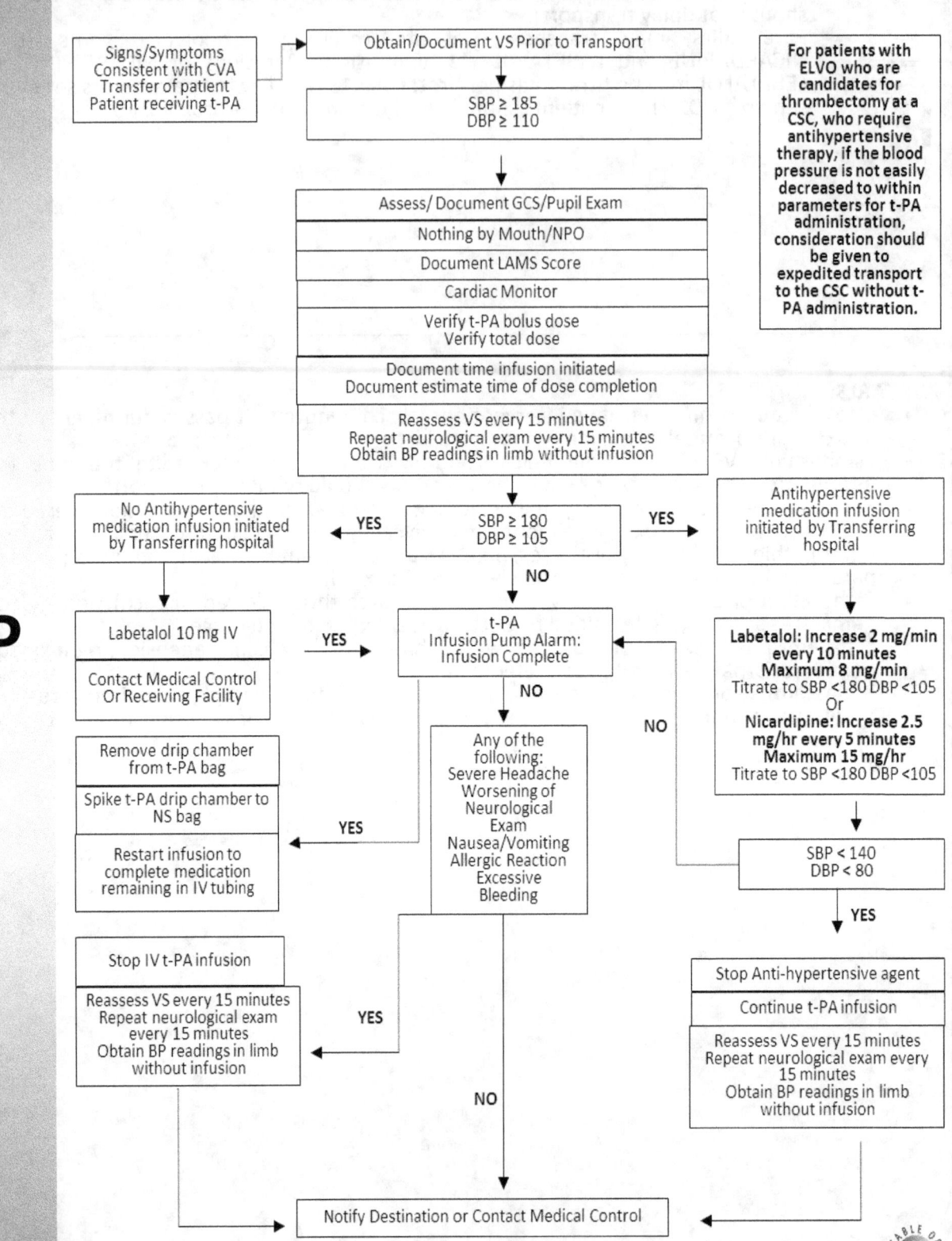

P

Signs/Symptoms Consistent with CVA Transfer of patient Patient receiving t-PA

Obtain/Document VS Prior to Transport

SBP ≥ 185
DBP ≥ 110

For patients with ELVO who are candidates for thrombectomy at a CSC, who require antihypertensive therapy, if the blood pressure is not easily decreased to within parameters for t-PA administration, consideration should be given to expedited transport to the CSC without t-PA administration.

Assess/ Document GCS/Pupil Exam

Nothing by Mouth/NPO

Document LAMS Score

Cardiac Monitor

Verify t-PA bolus dose
Verify total dose

Document time infusion initiated
Document estimate time of dose completion

Reassess VS every 15 minutes
Repeat neurological exam every 15 minutes
Obtain BP readings in limb without infusion

No Antihypertensive medication infusion initiated by Transferring hospital ← **YES** — SBP ≥ 180 DBP ≥ 105 — **YES** → Antihypertensive medication infusion initiated by Transferring hospital

NO

Labetalol 10 mg IV

Contact Medical Control Or Receiving Facility

YES →

t-PA
Infusion Pump Alarm:
Infusion Complete

NO

Remove drip chamber from t-PA bag

Spike t-PA drip chamber to NS bag

Restart infusion to complete medication remaining in IV tubing

YES ←

Any of the following:
Severe Headache
Worsening of Neurological Exam
Nausea/Vomiting
Allergic Reaction
Excessive Bleeding

**Labetalol: Increase 2 mg/min every 10 minutes
Maximum 8 mg/min**
Titrate to SBP <180 DBP <105
Or
**Nicardipine: Increase 2.5 mg/hr every 5 minutes
Maximum 15 mg/hr**
Titrate to SBP <180 DBP <105

NO

SBP < 140
DBP < 80

YES

Stop IV t-PA infusion

Reassess VS every 15 minutes
Repeat neurological exam every 15 minutes
Obtain BP readings in limb without infusion

YES ←

Stop Anti-hypertensive agent

Continue t-PA infusion

Reassess VS every 15 minutes
Repeat neurological exam every 15 minutes
Obtain BP readings in limb without infusion

NO

Notify Destination or Contact Medical Control

Recognition:
- Anterior epistaxis is identified by blood draining primarily from one or both nares.
- Posterior epistaxis is identified by the observation of blood draining into the posterior pharynx.

E

- Routine Patient Care.
- For patients with significant or multisystem trauma, manage per the appropriate protocol(s).
- Compress nostrils with direct pressure, tilt head forward, place patient in position of comfort.
- If bleeding is not controlled, have patient blow their nose, suction active bleeding as required, insufflate OXYMETAZOLINE 2 sprays to the affected nostril, and follow by direct pressure.
- Consider the need for airway management.
- If the patient is hypotensive, also manage per the age appropriate *Hemorrhagic Shock Protocol.*
- If bleeding is not controlled or the patient is hypotensive, transport the patient to the nearest appropriate Hospital Emergency Facility.

P

- For uncontrolled bleeding, consider placement of nasal packing following the *Nasal Packing Procedure Protocol*.
- All patients with nasal packing placed in the field will likely require prophylactic antibiotics and therefore require transport.

PEARLS:
- It may be difficult to quantify the amount of blood loss with epistaxis.
- 90% of nosebleeds are anterior in etiology, however posterior bleeding may be present. Evaluate for posterior by examining the posterior pharynx.
- Obtain medication history in all patients with epistaxis.
- Anticoagulants including warfarin (Coumadin), heparin, enoxaparin (Lovenox), dabigatran (Pradaxa), rivaroxban (Xarelto), and many over the counter headache relief powders may contribute to bleeding.
- Antiplatelet agents including ASA, clopidogrel (Plavix), ASA/dipyridamole (Aggrenox), and ticlopidine (Ticlid) may contribute to bleeding.
- Oxymetazoline is contraindicated in patients with known coronary artery disease or those with a diastolic blood pressure > 110 .
- Prolonged nasal packing has been associated with toxic shock syndrome.

Recognition:
- Body temperature ≥ 100.4 (f)/38 (c).

E

- Routine patient care.
- Institute passive cooling (remove excessive clothing/bundling).
- If no acetaminophen has been taken/administered in the last 4 hours, administer ACETAMINOPHEN 500-1000 mg PO.
- If acetaminophen has been taken/administered in the last 4 hours and the body temperature is ≥ 100.4 f)/38(c), administer IBUPROFEN 400-800 mg PO.
- If only IBUPROFEN has been administered in the last 6 hours and the body temperature is ≥ 100.4(f)/38(c), administer ACETAMINOPHEN 500-1000 mg PO.
- Exit to appropriate protocol(s)/transport the patient to nearest appropriate Hospital Emergency Facility.

PEARLS:
- Ascertain medication allergies prior to administering any medications. An allergy to non-steroidal anti-inflammatory medications (NSAIDs) is a contraindication to the administration of ibuprofen.
- Ibuprofen should not be administered to patients with preexisting renal disease/insufficiency or that are pregnant. Administer Ibuprofen with caution in patients with dehydration.
- Acetaminophen should not be administered to patients with liver disease.
- Consider whether elevated temperature is due to "fever" (suspected infection) vs. an environmental heat emergency.
- NSAIDs should not be administered in the setting of environmental heat emergencies.
- Rehydration with fluids will increase the patient's ability to sweat and may improve temperature control.
- The primary goal of treating fever is increasing patient comfort vs. normalization of body temperature. Absent neurologic injury/insult (traumatic brain injury, CVA, post cardiac arrest states) there is no evidence that fever worsens illness.

Recognition:
- Body temperature ≥ 100.4 (f)/38 (c).

E

- Routine patient care.
- Institute passive cooling (remove excessive clothing/bundling).
- If no acetaminophen has been taken/administered in the last 4 hours, administer ACETAMINOPHEN 15 mg/kg PO/PR (1000 mg max)
- If acetaminophen has been taken/administered in the last 4 hours and the body temperature is ≥ 100.4 (f)/38(c), administer IBUPROPHEN 10 mg/kg PO (800 mg max) if patient is ≥6 months of age.
- If only ibuprofen has been administered in the last 6 hours and the body temperature is ≥ 100.4 (f)/38(c), administer ACETAMINOPHEN 15 mg/kg mg PO/PR (max 1000 mg).
- Exit to appropriate protocol/transport the patient to nearest appropriate Hospital Emergency Facility.

PEARLS:
- All patients < 2 months of age with fever require a complete workup and should be transported to the hospital.
- Ascertain medication allergies prior to administering any medications. Allergy to non-steroidal anti-inflammatory medications (NSAIDs) is a contraindication to the administration of ibuprofen.
- Ibuprofen should not be administered to patients with preexisting renal disease/insufficiency. Administer Ibuprofen with caution in patients with dehydration.
- Acetaminophen should not be given to patients with liver disease.
- Consider whether elevated temperature is due to "fever" (suspected infection) vs. an environmental heat emergency.
- NSAIDs should not be administered in the setting of environmental heat emergencies.
- Rehydration with fluids will increase the patient's ability to sweat and may improve temperature control.
- The primary goal of treating fever is increasing patient comfort vs. normalization of body temperature. Absent neurologic injury/insult (traumatic brain injury, CVA, post cardiac arrest states) there is no evidence that fever worsens illness.

Recognition:
- Newly born infant meeting any of the following criteria:
  - Less than term gestation (<37 weeks)
  - Not crying/breathing or has a HR <100
  - Poor muscle tone
  - Labored breathing/gasping, or persistent (>5-10 min) central cyanosis

- Routine patient care.
- Perform the following within the first 60 seconds of delivery:
  - Warm the infant and maintain normothermia.
  - Position the infant to establish and maintain a patent airway.
  - Clear airway secretions by suctioning with a bulb syringe or suction catheter only if secretions are copious and/or obstructing the airway or positive pressure ventilation is required.
  - Stimulate the infant.
  - Assess breathing and heart rate (HR) [HR should be assessed by the use of 3 lead ECG monitoring if utilization of such is appropriate for provider level, alternative methods of determining HR when ECG monitoring is unavailable is by auscultation of the apical pulse or by palpating the base of the umbilical cord].

- If the infant's **HR is >100 bpm, but breathing is labored or there is persistent central cyanosis:**
  - Reposition and clear the airway as indicated.
  - Utilize pulse oximetry to assess oxygenation (the SpO_2 should be measured utilizing the right hand [pre-ductal]).
  - Provide supplemental oxygen if needed to achieve targeted preductal oxygen saturations as outlined in Table 1 below.
  - Monitor and reassess. If the infant becomes apneic, begins gasping or the HR decreases to <100, manage as below.

- If the infant is **apneic or gasping, or the HR is <100 bpm:**
  - Provide positive pressure ventilation (PPV) at a rate of 40-60 bpm.
  - In term infants (>37 weeks), initial PPV may be provided with room air.
  - In infants of <35 weeks gestation, initial PPV should be provided with low concentration OXYGEN (<30%). Titrate supplemental oxygen to achieve targeted preductal oxygen saturations in Table 1 below.
  - Continue to monitor HR and provide PPV until the HR >100 bpm.
  - Supplemental oxygen should be titrated down as soon as possible.

- If the **HR is <60 bpm:**
  - Provide PPV with supplemental OXYGEN at a rate of 40-60 bpm for a period of 30 seconds. If the HR increases to >60 bpm, but is <100 bpm, monitor HR and continue provide PPV until the HR is >100 bpm.
  - If the HR remains < 60 bpm after 30 seconds of PPV, provide external chest compressions following current AHA Guidelines for CPR and ECC (two thumb technique preferred, 3:1 ratio of compressions to ventilations with 90 compressions and 30 breaths to achieve approximately 120 events/minute). Infants requiring continued chest compressions should receive PPV with high concentration (FiO2 1.0) OXYGEN.
  - Reassess the HR every 60 seconds. If the heart rate fails to increase, check for adequate chest rise and take corrective actions as indicated.
  - Supplemental oxygen should be titrated down as soon as possible.
- Calculate and document 1 and 5 minute APGAR Scores (calculation of APGAR scores should not delay resuscitation) see Table 2 below.
- Transport patient to the nearest appropriate Hospital Emergency Facility.

E

P

- Provide advanced airway management (ETI/BIAD) as indicated per age appropriate *Airway Management Protocol.*
- Consider IV/UVC access as indicated by the clinical situation and response to therapy.
- For patients requiring continued chest compressions, EPINEPHRINE (1:10:000) 0.01-0.03 mg/kg IV, repeat every 3-5 min. In the event IV access is not yet established, consider EPINEPHRINE (1:10,000) 0.05-0.10/kg via ETT (IV administration if preferred).
- If hypovolemia is suspected (pale skin, poor perfusion, weak pulse, HR unresponsive to resuscitative measures), consider NORMAL SALINE 10 ml/kg over 5-10 minutes (may repeat X1).
- All critically ill neonates should have their bG determined. If hypoglycemia is present (bG <50 mg/dl), administer DEXTROSE 10% 2-4 ml/kg IV at 1 ml/min and recheck the bG.
- Perform needle thoracostomy for suspected tension pneumothorax..

Table 1: Targeted Preductal SpO$_2$

1 minute	60%-65%
2 minute	65%-70%
3 minute	70%-75%
4 minute	75%-80%
5 minute	80%-85%
10 minute	85%-95%

Table 2: APGAR Scoring System – Calculate and 1st and 5th Minute of Life

Criteria	2	1	0
Activity (muscle tone)	Active	Arms and legs flexed (weak, some movement)	Limp or flaccid
Pulse	>100	<100	Absent
Grimace	Cry, sneeze, cough, active movement	Grimace (some flexion of extremities)	No reflexes
Appearance	Completely pink	Body pink, extremities blue	Blue, pale
Respiration	Vigorous cry	Slow, irregular, or gasping, weak cry	Absent

PEARLS:

- Patients not meeting the above criteria should remain with the mother and be provided routine post-partum newborn infant care as outlined in the *Obstetrical Delivery-Labor Protocol.*
- Pulse oximetry is inaccurate in determining heart rate during the first few minutes of life and should not be used for this purpose.
- In uncompromised neonates, blood oxygen levels may not reach extra uterine levels until approximately 10 minutes after birth.
- Attaching the SpO2 probe to the neonate before connecting the probe to the monitor may facilitate more rapid signal acquisition.
- Preductal oxygen saturations are more representative of brain oxygenation.
- Peripheral cyanosis affects the hands and feet and is caused by peripheral vasoconstriction. It is a common benign condition in the newborn.
- Central cyanosis affects the mucous membranes, lips, skin, and nailbeds, should be considered pathological until proven otherwise.
- Initiating resuscitation with high oxygen concentrations (≥65%) is not recommended and to reduce the risks associated with hyperoxia, supplemental oxygen concentrations should be weaned as soon as possible.
- Infants born through meconium stained amniotic fluid who are vigorous with good respiratory effort and muscle tone may stay with the mother to receive the initial steps of newborn care.
- Routine intubation for tracheal suctioning in the presence of meconium staining is not recommended.
- When providing PPV, a manometer should be used to monitor the positive inspiratory pressure (PIP) delivered. An initial inflation pressure of 20 cmH20 may be effective, but ≥30-40 cmH20 may be required in some term infants without spontaneous breathing.
- Risk for hypoglycemia include prematurity, small for gestational age, infant of a diabetic mother, stress or sickness.
- A general rule of thumb for normal neonatal blood pressure is that the MAP should equal the gestational age in weeks.
- Rapid administration of large volumes of resuscitation fluids neonates has been associated with intraventricular hemorrhage. Resuscitation fluids should be administered over 5-10 minutes.
- Needle thoracostomy should be performed utilizing an 18, 20, or 22 gauge IV catheter (preferred) or a 23 or 25 gauge scalp vein (butterfly) needle.

- Routine patient care.
- Obtain history:
  - Length of gestation and due date
  - Time of onset of contractions, intensity, duration and interval
  - +/- rupture of membranes
  - Gravida/para status
  - Presence of vaginal bleeding
  - High risk factors
  - Past medical history/medications
  - Use of narcotics within last 4 hours
  - +/- Prenatal care
- Identify imminent delivery (+ rupture of membranes, contractions ≤ 2 min apart, crowning, urge to push or move bowels).
- If imminent delivery is not identified or there is evidence of complications (breech, prolapsed umbilical cord), initiate transport, manage complications per the *Obstetrical Complications Protocol.*

- If an **uncomplicated imminent delivery is identified, stay on scene** and immediately prepare to assist with delivery.
  - Position mother supine on flat surface, if possible.
  - Administer high flow oxygen.
  - Do not attempt to impair or delay delivery.
  - Support and control delivery of the head as it emerges.
  - Protect the perineum with gentle hand pressure.
  - After partial delivery of the head and prior to delivery of other fetal parts, suction the nasopharynx and oropharynx and perform a finger sweep of the fetal neck to check for the presence of a nuchal cord.
  - If a nuchal cord is identified, the cord should be disentangled by slipping the cord over the baby's head. If disentanglement of the cord is not possible, the cord should be double-clamped and cut prior to further delivery efforts.
  - As the shoulders emerge, gently guide the head and neck downward to deliver the anterior shoulder.
  - Support and gently lift the head and neck to deliver the posterior shoulder.
  - If delivery is not progressing due to suspected shoulder dystocia, manage per the *Obstetrical Complications Protocol.*
  - After delivery of the shoulders, the remainder of the fetus should deliver passively and with minimum difficulty.

- Provide routine post-partum infant care as follows:
  - Wrap in infant in blanket, maintain normal temperature.
  - Suction the mouth and nose only if signs of obstruction are present.
  - If the infant is not vigorous, stimulate by flicking the soles of the feet and/or rubbing the infant's back.
  - If the infant is not of term gestation, has poor tone, is not crying or breathing, is cyanotic or has a HR <100, double clamp and cut umbilical cord and manage per the *Neonatal Resuscitation Protocol.*
  - For infants not requiring resuscitation, delay clamping and cutting the umbilical for 30 sec following delivery. Clamp the cord 6″ from the infant's abdominal wall. The infant should be maintained at the level of the uterus until the cord is clamped.
  - Allow infant to nurse.
  - Document 1 and 5 min APGAR scores (see table 1).
  - Transport newborn infant in appropriately sized child safety seat.

E

E

- Provide maternal postpartum care:
  - Encourage the mother to deliver the placenta (placenta should deliver within 20-30 minutes). Do not pull on the umbilical cord to hasten delivery of the placenta. Once the placenta is delivered, note if it is intact and place in a plastic bag and transport it with the mother.
  - Inspect the perineum for tearing or bleeding; apply direct pressure with sanitary pads if indicated.
  - Manage postpartum hemorrhage (>500 ml estimated blood loss or blood loss with hemodynamic instability) per _Obstetrical Complications Protocol_

Table 1: APGAR Scoring System – Calculate and 1st and 5th Minute of Life

Criteria	2	1	0
Activity (muscle tone)	Active	Arms and legs flexed (weak, some movement)	Limp or flaccid
Pulse	>100	<100	Absent
Grimace	Cry, sneeze, cough, active movement	Grimace (some flexion of extremities)	No reflexes
Appearance	Completely pink	Body pink, extremities blue	Blue, pale
Respiration	Vigorous cry	Slow, irregular, or gasping, weak cry	Absent

PEARLS:
- Normal pregnancy is accompanied by higher heart rates and lower blood pressures. Shock will be manifested by signs of poor perfusion.
- Labor can take 8-12 hours, but as little as 5 minutes if high PARA (the higher the PARA, the shorter the labor is likely to be).
- High risk factors include: no prenatal care, drug use, teenage pregnancy, DM, HTN, cardiac disease, prior breech or C section, preeclampsia, twins.
- Newborns are prone to hypothermia which may lead to hypoglycemia, hypoxia, and lethargy. Aggressive warming techniques should be initiated including drying, swaddling, and warm blankets.

Recognition:
- Prolapsed umbilical cord: umbilical cord precedes the fetus.
- Shoulder dystocia: failure of the fetal shoulder to deliver shortly after delivery of the head.
- Malpresentation (Breech): presentation of the fetal buttocks or legs.
- 3rd trimester bleeding: vaginal bleeding occurring ≥ 28 weeks of gestation.
- Postpartum hemorrhage: >500 ml estimated blood loss or blood loss with hemodynamic instability.
- Preterm labor: onset of labor/contractions prior to the 37th week of gestation.

E

- Routine patient care.
- Do not delay transport for patients with obstetrical complications. Provide early notification to the receiving facility.

> **The following actions may not be feasible in every case, nor may every obstetrical complication be anticipated or effectively managed in the field. These should be considered "best advice" for rare, difficult scenarios.**

- Administer high-flow oxygen.
- For **prolapsed umbilical cord**:
  - Discourage pushing by the mother.
  - Place the patient in the knee chest or Trendelenberg position.
  - Place a gloved hand in the mother's vagina and decompress the umbilical cord by elevating the presenting fetal part off of the cord.
  - Feel for cord pulsation and monitor for fetal bradycardia (HR <120).
  - Keep the exposed cord moist and warm. Do not replace it.

- For **shoulder dystocia**:
  - Discourage pushing by the mother.
  - Support the baby's head, do not pull on it. Suction the nasopharynx and oropharynx as needed.
  - Place the patient in an extreme the knee chest position (McRobert's maneuver) and apply gentle suprapubic pressure with an open hand.
  - If the above method is unsuccessful, consider rolling the patient from her existing position to the all-fours position. Often, the shoulder will dislodge during the act of turning, so that this movement alone may be sufficient to dislodge the impaction. In addition, once the position change is completed, gravitational forces may aid in the disimpaction of the fetal shoulders.

- For **malpresentation (breech) delivery:**
  - <u>Transport</u> unless delivery is imminent.
  - Support presenting parts (never attempt to pull the presenting parts).
  - If the legs have delivered, gently elevate the trunk and legs to aid delivery of the head.
  - If the head is not delivered within 30 seconds of the legs, place two fingers into the vagina to locate the infant's mouth. Press the vaginal wall away from the infant's mouth to maintain the fetal airway.
  - Apply gentle pressure to the uterine fundus.

- For **3rd trimester bleeding**:
  - Suspect placenta previa (placenta is implanted in the lower uterine segment) or placental abruption (placenta is separated from the uterine wall before delivery).
  - Do not perform a digital examination.
  - Place the patient in the left lateral recumbent position.
  - Manage hypotension or shock per the *Hemorrhagic Shock Protocol*
- If the infant is delivered, provide postpartum care per the *Obstetrical Delivery-Labor Protocol.*

E

- For **postpartum hemorrhage**:
  - o Provide vigorous fundal massage until the uterus is firm.
  - o If possible, initiate infant nursing.
  - o Treat hypotension or shock per age appropriate *Hemorrhagic Shock Protocol.*
- For **seizures**, manage per the age appropriate *Seizure Protocol.*
- For **cardiac arrest in the pregnant patient** (regardless of etiology):
  - o Follow appropriate medical *Cardiac Arrest Protocol(s)* or *Traumatic Cardiac Arrest Protocol.*
  - o For patients ≥ 20 weeks gestation or if the fundus is palpable at or above the level of the umbilicus, apply left lateral uterine displacement (LUD) with the patient in the supine position to decrease aortocaval compression. LUD should be maintained during CPR. If ROSC is achieved, the patient should be placed in the left lateral decubitus position or, if the patient is on a long spine board, the board can be wedged 15 degrees to the left.
  - o IV access should be obtained above the diaphragm.
  - o Consider rapid transport to the nearest Hospital Emergency Facility for possible peri-mortem cesarean delivery (PMCD). Provide early notification to the receiving facility.
- Transport patient to the nearest appropriate Hospital Emergency Facility.

P

- For **preterm labor**, consider:
  - o NORMAL SALINE 20 ml/kg IV (may repeat X1).
  - o MAGNESIUM SULFATE 4 gm IV over 20 min.
  - o ALBUTEROL 2.5-5 mg via SVN.
- For **postpartum hemorrhage**, after delivery of the placenta, PITOCIN 10 -20U IM (if not previously administered) followed by PITOCIN 20 units/1 L NS/LR at 40 mU/min.
- For **seizures** in the pregnant patient (>20 weeks gestation), MAGNESIUM SULFATE 4 gm/IV over 10 min (may repeat 2 gm in 5 min).

PEARLS:
- Shoulder dystocia usually occurs when the fetal anterior shoulder impacts against the maternal symphysis following delivery of the vertex. Less commonly, shoulder dystocia results from impaction of the posterior shoulder on the sacral promontory.
- Placental abruption may present with mild, dark vaginal bleeding with abdominal pain/ uterine tenderness to massive hemorrhage and fetal demise. Because hemorrhage may occur into the abdominal/pelvic cavity, shock can develop despite relatively little visible (vaginal) bleeding.
- Placenta previa is usually associated with painless, bright red vaginal bleeding and a non-tender, soft uterus. Unlike the case with abruption, placenta previa related hemorrhage passes through the vaginal outlet and is not occult.
- Maternal hypotension is defined as a SBP < 100 or < 80% of baseline.
- Maternal hypotension can result in decreased placental perfusion.
- In the setting of pregnancy, HTN is defined as a SBP > 140 or a DBP > 90, or a relative increase of 30 systolic and 20 diastolic from the patient's normal (pre-pregnancy) BP.
- Severe headache, visual changes, edema, or RUQ pain may indicate preeclampsia.
- Eclamptic seizures may occur up to two months post-partum.
- For the pregnant patient in cardiopulmonary arrest, best outcomes for the mother and the fetus are achieved by aggressive maternal resuscitation.
- Perform high quality CPR and follow appropriate *Cardiac Arrest Protocols*.
- The most common causes of maternal cardiopulmonary arrest are hemorrhage, cardiovascular disease (MI, Aortic dissection, myocarditis), pulmonary embolism, amniotic fluid embolism, eclampsia and sepsis.
- Aortocaval compression can occur at 20 weeks gestation.
- At 20 weeks gestation the fundus is palpable at or above the umbilicus.

Recognition:
- Generalized or grand mal: loss of consciousness accompanied by bilateral tonic-clonic activity.
- Petit mal: "absence" period in which patients are not fluently conversant and have diminished neurological status.
- Focal or partial: seizure activity that is localized to one part of the body.

E

- Routine patient care.
- Perform blood glucose (bG) analysis, manage per the age appropriate *Diabetic Emergencies Protocol* if indicated.
- Loosen any constrictive clothing, protect patient and providers.
- If the patient has been prescribed DIAZEPAM rectal gel or MIDAZOLAM (IN or via auto-injector) assist the family or caregiver with administration following the prescribed instructions.
- If the patient has a vagal nerve stimulator (VNS) device, assist family members or caregiver with use if needed. The VNS is activated by passing a magnet closely over the device. This may be repeated every 3-5 min for a total of 3 times.
- Transport the patient to the nearest appropriate Hospital Emergency Facility.

A C

- For **active generalized or focal seizures in the patient without IV access**: MIDAZOLAM 10 mg IM or 2 mg IN.
- For **active generalized or focal seizures in the patient with IV access** or the patient with seizure activity refractory to MIDAZOLAM as above:
  - ○ LORAZEPAM 4 mg IV (repeat 2 mg every 3-5 min to a max of 10 mg) **or**
  - ○ MIDAZOLAM 2.5 mg IV [5 mg IM or 2 mg IN] (repeat 2 mg every 3-5 min to a max of 20 mg) **or**
  - ○ DIAZEPAM 5 mg IV (repeat every 3-5 min to a max of 20 mg).

P

- For **active generalized or focal seizures in the patient without IV access**: MIDAZOLAM 10 mg IM or 2 mg IN.
- For **active generalized or focal seizures in the patient with IV access** or the patient with seizure activity refractory to MIDAZOLAM as above:
  - ○ LORAZEPAM 4 mg IV (repeat 2 mg every 3-5 min to a max of 10 mg) **or**
  - ○ MIDAZOLAM 2.5 mg IV [5 mg IM or 2 mg IN] (repeat 2 mg every 3-5 min to a max of 20 mg) **or**
  - ○ DIAZEPAM 5 mg IV (repeat every 3-5 min to a max of 20 mg).
- For seizures refractory to benzodiazepines, consider LEVETIRACETAM 20 mg/kg IV over 15 min (may repeat once).
- For seizures in a <u>known or suspected pregnant patient</u> (> 20 weeks gestation), manage per the *Obstetrical Complications Protocol*.

PEARLS:

- Recommended exam: mental status, HEENT, cardiac, pulmonary, extremities, neurologic.
- Do not delay medication administration to obtain vascular access. The IM/IN routes are preferred if there is no immediate IV access at the time of EMS contact.
- Lorazepam is the first-line antiepileptic of choice if IV/IO access is available.
- Diazepam has the longest ½ life and the least desirable choice when other benzodiazepines are available. Diazepam may be administered rectally.
- Respiratory depression may occur after the administration of benzodiazepines, monitor the patient closely and utilize waveform capnography if available.
- Phenobarbital has additive respiratory depressant effects.
- Status epilepticus (SE) is defined as two or more successive seizures without a period of consciousness or recovery, or one prolonged seizure lasting > 5 minutes. SE is a true emergency requiring rapid airway control and treatment.
- For seizures in the pregnant patient, magnesium is the first line medication.

Recognition:
- Generalized or grand mal: loss of consciousness accompanied by bilateral tonic-clonic activity.
- Petit mal: "absence" period in which patients are not fluently conversant and have diminished neurological status.
- Focal or partial : seizure activity that is localized to one part of the body.

E

- Routine patient care.
- Perform blood glucose (bG) analysis, manage per age appropriate *Diabetic Emergencies Protocol* if indicated.
- Loosen any constrictive clothing, protect patient and providers.
- For the patient with active generalized or focal seizures with a prescription for DIAZEPAM rectal gel or MIDAZOLAM (IN or via Auto-injector) assist the family or caregiver with administration following the prescribed instructions.
- If the patient has a vagal nerve stimulator (VNS) device, assist family members or caregiver with use if needed. The VNS is activated by passing a magnet closely over the device. This may be repeated every 3-5 min for a total of 3 times.
- For suspected **febrile seizures** (generalized seizures with no seizure history [except previous febrile seizures] in the setting of any grade fever, with an otherwise normal neurologic examination):
  - Obtain body temperature.
  - Passive cooling measures (undress patient).
  - ACETAMINOPHEN 15 mg/kg suppository PR if the temperature is ≥ 100.4.
- Any seizure lasting > 5 min requires treatment with anti-seizure medications (benzodiazepines), consider rapid ALS intercept if available or transport the patient without delay to the nearest appropriate Hospital Emergency Facility.

A
C

- For **active generalized or focal seizures:**
  - MIDAZOLAM 0.2 mg/kg IM/IN or 0.1 mg/kg IV [4 mg max single dose via any route)] (may repeat every 5 min to a max of 1 mg/kg) **or**
  - LORAZEPAM 0.1 mg/kg IV [4 mg max single dose] (may repeat every 5 min to a max of 0.5 mg/kg) **or**
  - DIAZEPAM 0.1 mg/kg IV [5 mg max single dose] or 0.5 mg/kg PR [20 mg max single dose] (may repeat every five min to a max of 1 mg/kg).

P

- For **active generalized or focal seizures**:
  - MIDAZOLAM 0.2 mg/kg IM/IN or 0.1 mg/kg IV [4 mg max single dose via any route)] (may repeat every 5 min to a max of 1 mg/kg) **or**
  - LORAZEPAM 0.1 mg/kg IV [4 mg max single dose] (may repeat every 5 min to a max of 0.5 mg/kg) **or**
  - DIAZEPAM 0.1 mg/kg IV [5 mg max single dose] or 0.5 mg/kg PR [20 mg max single dose] (may repeat every five min to a max of 1 mg/kg).
- For **persistent seizures**, consider, if available, PHENOBARBITAL 20 mg/kg IV at rate of < 50 mg/min (may repeat 5 mg/kg IV every 5 min until seizure activity is terminated) **or** LEVETIRACETAM 20 mg/kg IV over 15 min (may repeat once).

PEARLS:

- Recommended exam: mental status, HEENT, cardiac, pulmonary, extremities, neurologic.
- Do not delay medication administration to obtain vascular access. The IM/IN route is preferred if no immediate IV access at the time of EMS contact.
- Lorazepam is the first-line antiepileptic of choice if IV/IO access is available.
- Diazepam has the longest ½ life and the least desirable choice when other benzodiazepines are available. Diazepam may be administered rectally.
- Respiratory depression may occur after the administration of benzodiazepines, monitor the patient closely and utilize waveform capnography if available.
- Phenobarbital has additive respiratory depressant effects.
- Status epilepticus (SE) is defined as two or more successive seizures without a period of consciousness or recovery, or one prolonged seizure lasting > 5 minutes. SE is a true emergency requiring rapid airway control and treatment.
- A focal to generalized seizure starts in one area and generalizes to the whole body.
- The area where the seizure starts is typically the last to resolve. Knowing this focality is helpful.

Recognition:
- This protocol applies to undifferentiated, hypovolemic, obstructive, cardiogenic, and neurogenic shock.
- Restlessness, confusion, pale and cool clammy skin.
- Weak, rapid pulse, hypotension (SBP <90 or MAP <65) or shock index (HR/SBP) ≥0.6.
- Hypovolemic: history of volume loss (vomiting, diarrhea).
- Obstructive: history of chest trauma, JVD, absent breath sounds, narrow pulse pressure, history of DVT, risk factors for pulmonary embolus.
- Cardiogenic: STEMI, heart failure, dysrhythmias, bradycardia or tachycardia.
- Neurogenic: spinal cord injury with hypotension, warm & dry skin with a normal or bradycardic heart rate.

E
- Routine patient care.
- Manage suspected **obstructive shock** per protocols specific to etiology (e.g. _Multiple Trauma Protocol_ for suspected tension pneumothorax).
- If tolerated, maintain patient in a supine position.
- Maintain and promote normothermia by use of and increasing the ambient temperature if possible.
- Transport the patient to the nearest appropriate Hospital Emergency Facility.

A
C
- LACTATED RINGER'S 500 ml IV, repeat to achieve a SBP of ≥100 or MAP of ≥65 (max 2L).
- If **adrenal insufficiency** is suspected, also manage per the _Adrenal Insufficiency Protocol_.

P
- For **undifferentiated and hypovolemic shock:**
  - LACTATED RINGER'S 500 ml IV, repeat to achieve a SBP of ≥100 or MAP of ≥65 (max 2L).
  - For persistent hypotension following the administration of 2L of IVF, consider NOREPINEPHRINE 2-20 mcg/minute **or** DOPAMINE HCL 2-20 mcg/kg/min **or** PHENYLEPHRINE 100 mcg every 10 min [max dose 500 mcg] or PHENYLEPHRINE 10-180 mcg/min Ito achieve a SBP of ≥100 or MAP ≥65.
- For **post arrest hypotension**:
  - NOREPINEPHRINE 2-20 mcg/minute **or** DOPAMINE HCL 2-20 mcg/kg/min **or** PHENYLEPHRINE 100 mcg every 10 min [max dose 500 mcg] **or** PHENYLEPHRINE 10-180 mcg/min to achieve a SBP of ≥100 or MAP ≥65.
- For **cardiogenic shock:**
  - LACTATED RINGER'S 250 ml IV, (do not administer IVF if there is shortness of breath or evidence of pulmonary edema/CHF), repeat X1 if needed to achieve SBP ≥100 or MAP ≥65.
  - NOREPINEPHRINE 2-20 mcg/min (preferred) **or** DOPAMINE HCL 2-20 mcg/kg/min to achieve a SBP ≥100 or MAP ≥65.
- For **neurogenic shock:**
  - LACTATED RINGER'S 500 ml IV, repeat to achieve a SBP of ≥100 or MAP of ≥65.
  - For persistent hypotension following the administration of 2L of IVF, PHENYLEPHRINE 10-180 mcg/min **or** NOREPINEPHRINE 2-20 mcg/min **or** DOPAMINE HCL 2-20 mcg/kg/min to achieve a SBP of ≥100 or MAP ≥65.
- If **adrenal insufficiency** is suspected, also manage per the _Adrenal Insufficiency Protocol._

PEARLS:
- Shock is a state in which inadequate tissue perfusion impairs cellular metabolism.
- Shock may be present with a normal blood pressure.
- Blood pressure and heart rate individually are unreliable determinants of hypovolemic shock.
- The systolic blood pressure must be evaluated in the context of the patient's normal blood pressure. Patients with a history of long standing hypertension often require a higher MAP to maintain perfusion pressures.
- Consider all possible etiologies of undifferentiated hypotension/shock and treat accordingly per appropriate protocols.
- Patients at increased risk for adrenal insufficiency (AI) include those with HIV/AIDS, a history of chronic steroid use (current, recently discontinued or remote), Addison disease, or dehydration.
- The use of the Trendelenburg position (head down, feet elevated) should be avoided.

- The use of push-dose phenylephrine allows for the rapid correction of acute hypotension and as bridge to the administration of a vasopressor via IV infusion.

Recognition:
- This protocol applies to undifferentiated, hypovolemic, obstructive, cardiogenic, and neurogenic shock.
- Restlessness, confusion, pale and cool clammy skin, prolonged capillary refill (CRT)
- Weak, rapid pulse, hypotension (SBP <70 + [age in years X 2] mmHg) [Neonates SBP <60 mmHg, 1 mo to 1 yr SBP <70 mmHg].
- Hypovolemic: history of volume loss (vomiting, diarrhea).
- Obstructive: history of chest trauma, JVD, absent breath sounds, narrow pulse pressure, history of DVT, risk factors for pulmonary embolus.
- Cardiogenic: STEMI, heart failure, dysrhythmias.
- Neurogenic: spinal cord injury with hypotension, warm & dry skin with a normal or bradycardic heart rate.

E

- Routine patient care.
- Treat suspected obstructive shock per age appropriate protocol specific to etiology (e.g. *Multiple Trauma Protocol* for suspected tension pneumothorax).
- If tolerated, maintain patient in a supine position.
- Maintain and promote normothermia by use of blankets and increasing the ambient temperature if possible.
- Transport to the nearest appropriate Hospital Emergency Facility, consider transport to a Pediatric Specialty Care Hospital.

A
C

- LACTATED RINGER'S 20 ml/kg IV bolus, repeat to achieve age appropriate BP (max 60 ml/kg).
- If **adrenal insufficiency** is suspected, also manage per the *Adrenal Insufficiency Protocol.*

P

- For **undifferentiated and hypovolemic shock**:
  - LACTATED RINGER'S 20 ml/kg IV, repeat to achieve age appropriate BP (max 60 ml/kg).
  - For persistent hypotension following the administration of 60 ml/kg of IVF, consider NOREPINEPHRINE 0.1-2 mcg/kg/min **or** EPINEPHRINE 0.1-1 mcg/kg/min **or** PHENYLEPHRINE 5 mcg/kg [max single dose 100 mcg] every 10 min [max total dose 500 mcg]. Titrate vasopressors to achieve age appropriate BP.
- For **cardiogenic shock**, consider DOPAMINE 2-20 mcg/kg/min titrated to achieve age appropriate BP.
- For **neurogenic shock**:
  - LACTATED RINGER'S 20 ml/kg IV, repeat to achieve age appropriate BP [max 60 ml/kg].
  - For persistent hypotension following the administration of 60 ml/kg of IVF, consider NOREPINEPHRINE 0.1-2 mcg/kg/min **or** EPINEPHRINE 0.1-1 mcg/kg/min. Titrate vasopressors to achieve age appropriate BP.
- If **adrenal insufficiency** is suspected, also manage per the *Adrenal Insufficiency Protocol.*

PEARLS:

- Shock is a state in which inadequate tissue perfusion impairs cellular metabolism.
- Shock may be present with a normal blood pressure.
- Blood pressure and heart rate individually are unreliable determinants of hypovolemic shock.
- The systolic blood pressure must be evaluated in the context of the patient's normal blood pressure. Patients with a history of long standing hypertension often require a higher MAP to maintain perfusion pressures.
- Consider all possible etiologies of undifferentiated hypotension/shock and treat accordingly per appropriate protocol.
- Patients at increased risk for adrenal insufficiency (AI) include those with HIV/AIDS, a history of chronic steroid use (current, recently discontinued or remote), Addison disease, or dehydration.
- The use of the Trendelenburg position (head down, feet elevated) should be avoided.

Recognition:
- Restlessness, confusion, pale and cool clammy skin.
- Weak, rapid pulse, hypotension (SBP <90 mmHg or MAP <65 mmHg) or shock index (HR/SBP) ≥0.6.
- External hemorrhage or overt or suspected occult internal hemorrhage (blunt trauma, GI bleeding, vaginal bleeding, suspected ruptured ectopic pregnancy, ruptured AAA).

E
- Routine patient care.
- Control external hemorrhage following the _External Hemorrhage Control Protocol._
- If tolerated, maintain patient in a supine position.
- Maintain and promote normothermia by use of blankets, heat reflective shield, and increasing the ambient temperature if possible.
- Transport to the nearest appropriate Hospital Emergency Facility.

A
C
- For **hemorrhagic shock related to external hemorrhage:**
  - Once hemorrhage control is achieved, LACTATED RINGER'S 500 ml IV, repeat to achieve a SBP ≥90 or MAP ≥65 (max 2L).
  - For patients with penetrating torso trauma and NO evidence of traumatic brain injury, LACTATED RINGER'S 250 ml IV, repeat to achieve a palpable radial pulse **or** a SBP ~80 or MAP of ~65 **or** normal mental status (max 2L).
  - For patients with penetrating torso trauma with evidence of traumatic brain injury, LACTATED RINGER'S 250-500 ml IV, repeat to achieve a SBP ≥90 or MAP ≥65 (max 2L).
- For **hemorrhagic shock related to suspected internal hemorrhage,** LACTATED RINGER'S 500 ml IV, repeat to achieve a SBP ≥90 or MAP ≥65 (max 2L).
- If **adrenal insufficiency** is suspected**,** manage per the _Adrenal Insufficiency Protocol_.

P
- For **hemorrhagic shock related to external hemorrhage**:
  - Once hemorrhage control is achieved, LACTATED RINGER'S 500 ml IV, repeat as needed to achieve a SBP ≥90 or MAP ≥65 (max 2L).
  - For patients with penetrating torso trauma and NO evidence of traumatic brain injury, LACTATED RINGER'S 250 ml IV, repeat to achieve a palpable radial pulse **or** a SBP ~80 or MAP ≥65 **or** normal mental status (max 2L).
  - For patients with penetrating torso trauma with evidence of traumatic brain injury, LACTATED RINGER'S 250-500 ml IV, repeat to achieve a SBP ≥90 or MAP ≥65 (max 2L).
  - For persistent hypotension following the administration of 2L of IVF, consider NOREPINEPHRINE 2-20 mcg/min **or** PHENYLEPHRINE 100 mcg IV every 10 min [max dose 500 mcg] to achieve a SBP of ≥90 or MAP ≥65. Vasopressors should only be considered AFTER the administration of 2L IVF and in the periarrest patient.
- For patients with blunt or penetrating trauma (including extremity trauma) with signs of significant hemorrhage (SBP < 90 mm Hg, HR > 110 BPM); or who are considered in paramedic judgment to be at high risk of significant hemorrhage (external or internal), administer TRANEXAMIC ACID 1gm/100 ml NS IV as soon as possible, but not >3 hours after injury.

- For **hemorrhagic shock related to internal hemorrhage:**
  - LACTATED RINGER'S 500 ml IV, repeat to achieve a SBP ≥90 or MAP ≥65 (max 2L).
  - For persistent hypotension following the administration of 2L of IVF, consider NOREPINEPHRINE 2-20 mcg/min **or** PHENYLEPHRINE 100 mcg IV every 10 min [max dose 500 mcg] to achieve a SBP of ≥90 or MAP ≥65. <u>Vasopressors should only be considered AFTER the administration of 2L IVF and in the periarrest patient.</u>
- If **adrenal insufficiency** is suspected, also manage per the *Adrenal Insufficiency Protocol.*

P

PEARLS:

- Shock is a state in which inadequate tissue perfusion impairs cellular metabolism.
- Shock may be present with a normal blood pressure.
- Blood pressure and heart rate individually are unreliable determinants of hypovolemic shock.
- The systolic blood pressure must be evaluated in the context of the patient's normal blood pressure. Patients with a history of long standing hypertension often require a higher MAP to maintain perfusion pressures.
- Permissive hypotension (hypotensive resuscitation) is indicated in patients with penetrating torso trauma.
- Consider all possible etiologies of hypotension/shock and treat accordingly per the appropriate protocol.
- Etiologies of internal hemorrhage include GI or vaginal bleeding, suspected ruptured ectopic pregnancy or AAA and blunt trauma.
- <u>Vasopressors are indicated as a bridge to surgical intervention in hemorrhagic shock only after adequate volume resuscitation and in the periarrest patient</u>.
- Patients at increased risk for adrenal insufficiency (AI) include those with HIV/AIDS, a history of chronic steroid use (current, recently discontinued or remote), Addison's disease, congenital adrenal hyperplasia (CAH) or dehydration.
- The use of the Trendelenburg position (head down, feet elevated) should be avoided.
- If tranexamic acid is administered, it is imperative that the receiving facility acknowledges the time of administration (there is a time window for the administration of a 2nd dose).

Recognition:
- Compensated: normal blood pressure, decreased distal pulses, delayed CRT, tachycardia, cool extremities, tachypnea, restlessness, +/- altered mental status.
- Uncompensated: Hypotension (SBP < 70 + [age in years X 2] mmHg.
- External hemorrhage or overt or suspected occult internal hemorrhage (blunt trauma, GI bleeding).
- For newborns the lower limit for SBP is <60, for children 1 mo to 1 yr the lower limit for SBP is <70.

E
- Routine patient care.
- Control external hemorrhage following the *External Hemorrhage Control Protocol*.
- If tolerated, maintain patient in a supine position.
- Maintain and promote normothermia by use of blankets and increasing the ambient temperature if possible.
- Transport to the nearest appropriate Hospital Emergency Facility, consider transporting to a Pediatric Specialty Care Hospital .

A C
- LACTATED RINGER'S 20 ml/kg, repeat to achieve age appropriate BP (max 60 ml/kg).
- If **adrenal insufficiency** is suspected, also manage per the *Adrenal Insufficiency Protocol.*

P
- LACTATED RINGER'S 20 ml/kg, repeat to achieve age appropriate BP (max 60 ml/kg).
- For patients with blunt or penetrating trauma (including extremity trauma) with signs of significant hemorrhage (age specific hypotension, tachycardia); or who are considered in paramedic judgment to be at high risk of significant hemorrhage (external or internal), administer TRANEXAMIC ACID as soon as possible, but not >3 hours after injury as follows: for patients < 12 yo, administer 15 mg/kg (max 1 gm) in 100 ml NS IV and for patients >12 yo, administer 1 gm/100 ml NS.
- For persistent hypotension following the administration of 60 ml/kg of IV fluids, consider NOREPINEPHRINE 0.1-2 mcg/kg/min IV **or** EPINEPHRINE 0.1-1 mcg/kg/min **or** PHENYLEPHRINE 5 mcg/kg IV [max single dose 100 mcg] every 10 min [max total dose 500 mcg].   Titrate vasopressors to achieve age appropriate BP.
- If **adrenal insufficiency** is suspected, manage per the *Adrenal Insufficiency Protocol.*

PEARLS:

- Shock is a state in which inadequate tissue perfusion impairs cellular metabolism.
- Shock may be present with a normal blood pressure.
- Children maintain BP and cardiac output by increasing HR and vasoconstriction, even with the loss of significant volume.
- A child can lose up to 1/3 of his/her blood volume before a significant drop in BP occurs.
- Pediatric hypotension should be viewed as a "pre-arrest state".
- Use of an appropriate size BP cuff is crucial as use of an incorrectly sized cuff will lead to inaccurate measurement of the BP (i.e. too small = falsely ↑BP, too large = falsely ↓ BP).
- Possible etiologies of internal hemorrhage include GI or vaginal bleeding, suspected rupture ectopic pregnancy or AAA and blunt trauma.
- Consider all possible etiologies of undifferentiated hypotension/shock and treat accordingly per the appropriate protocol.
- <u>Vasopressors are indicated as a bridge to surgical intervention in hemorrhagic shock only after adequate volume resuscitation and in the periarrest patient</u>.
- The use of the Trendelenburg position (head down, feet elevated) should be avoided.
- Patients at increased risk for adrenal insufficiency (AI) include those with HIV/AIDS, a history of chronic steroid use (current, recently discontinued or remote), Addison's disease, congenital adrenal hyperplasia (CAH) or dehydration.

Recognition:
- Sepsis is defined as life threatening organ dysfunction caused by a dysregulated host response to infection (the body's response to an infection injures its own tissues and organs). It is identified when a patient with a known infection (pneumonia, urinary tract, intra-abdominal, biliary, skin/soft tissue etc.) or with findings suggestive of infection (fever [T 100.4F/38 C], cough, presence of a wound with signs of infection etc.) has ≥2 of the following:
  1. Respiratory rate ≥22/min
  2. Acutely altered mentation (AMS) or increasing mental status change with previously AMS
  3. Systolic BP ≤100
- Septic shock is identified when a patient with sepsis, despite adequate fluid resuscitation requires vasopressors to maintain a SBP ≥100 or a MAP ≥65 and has a serum (fingertip) lactate >2 mmol/L or an EtCO2 ≤25 mmHg.

E
- Routine patient care.
- If tolerated, maintain patient in a supine position.
- Provide early notification to destination facility and declare a **SEPSIS ALERT.**
- Transport the patient to the nearest appropriate Hospital Emergency Facility.

A
C
- LACTATED RINGER'S 500 ml IV, repeat to achieve a SBP ≥100 or MAP of ≥65 (max 3L).
- If **adrenal insufficiency** is suspected, manage per the _Adrenal Insufficiency Protocol._

P
- If available, obtain fingertip lactate measurement.
- LACTATED RINGER'S 500 ml IV bolus, repeat to achieve a SBP ≥100 or MAP of ≥65 (max 3L)
- For hypotension refractory to IVF, consider NOREPINEPHRINE 2-20 mcg/min (preferred) **or** DOPAMINE HCL 2-20 mcg/kg/min **or** PHENYLEPHRINE 100 mcg IV every 10 min [max dose 500 mcg] to achieve a SBP of ≥100 or MAP ≥65.
- If **adrenal insufficiency** is suspected, management per the _Adrenal Insufficiency Protocol._

PEARLS:
- Shock may be present with a normal blood pressure.
- The qSOFA (quick sequential organ failure score) [MS, RR, SBP] is an abbreviated version of a larger scoring system (SOFA) used to identify organ failure.
- Assess for signs of pulmonary edema, especially patients with a known history of CHF, or ESRD on dialysis, stop fluid administration if pulmonary edema develops.
- The systolic blood pressure must be evaluated in the context of the patient's normal blood pressure. Patients with a history of long standing hypertension often require a higher MAP to maintain perfusion pressures.
- EtCO2 correlates to serum lactate levels. An EtCO2 of ≤25 mmHg is strongly correlated to a serum lactate of 4.0 mm/L.
- Septic shock has a 50% mortality rate and must be treated aggressively. Early goal directed therapy consisting of IV fluid administration and early antibiotics reduces mortality in septic patients.
- Common sites/sources for infection include the urinary tract, lungs, skin, GI tract and indwelling catheters and devices.
- Patients with sepsis are at an increased risk for acute lung injury (ALI) related to positive pressure ventilation. If positive pressure ventilation is required, avoid excessive tidal volumes.
- Patients at increased risk for adrenal insufficiency (AI) include those with HIV/AIDS, a history of chronic steroid use (current, recently discontinued or remote), Addison's disease, congenital adrenal hyperplasia or dehydration.

Recognition:
- Suspected or known infectious process **and** temperature abnormality (- >38.5°C or <36.0°C) **and a** heart rate greater than normal for age (see Table 1 below) **and one of the following:**
  1. Mental status abnormality (includes anxiety, restlessness, agitation, irritability, inappropriate crying, drowsiness, confusion, lethargy, obtundation).
  2. Perfusion abnormality (mottled or cool extremities, capillary refill time (CRT) <1 sec (flash) or > 3 sec, warm extremities, bounding pulse, SBP < 70 + 2X age in years)*.
  3. High risk condition (<56 days of life, BMT or solid organ x-plant, immune compromise, asplenia, sickle cell disease, malignancy).
  4. EtCO2 <25 mmHg.
  5. Finger stick lactate level >4 mmol/L.
- For newborns the lower limit for SBP is <60, for children 1 mo to 1 yr the lower limit for SBP is <70.

Table 1

Age	Tachycardia
1 mo - 1 yr	> 180
2 - 5 yr	> 140
6 - 12 yr	> 130
13-18 yr	> 120

E

- Routine patient care.
- If tolerated, maintain patient in a supine position.
- Perform blood glucose analysis, manage per age appropriate _Diabetic Emergencies Protocol_ as indicated.
- Provide early notification to destination facility and declare a **SEPSIS ALERT.**
- Transport to the nearest appropriate Hospital Emergency Facility, consider transporting to a Pediatric Specialty Care hospital.

A C

- LACTATED RINGER'S 20 ml/kg IV, repeat to achieve age appropriate BP (max 60ml/kg). Do not administer IV fluids if signs of fluid over load (crackles) develop.
- If **adrenal insufficiency** is suspected, manage per the _Adrenal Insufficiency Protocol._

P

- If available, obtain fingertip lactate measurement.
- LACTATED RINGER'S 20 ml/kg IV, repeat to achieve age appropriate BP (max 60ml/kg). Do not administer IV fluids if signs of fluid over load (crackles) develop.
- For patients with **cold shock** and continued hypotension or other evidence of hypoperfusion, consider EPINEPHRINE 0.1-1.0 mcg/kg/min titrated to achieve age appropriate BP.
- For patients with **warm shock** and continued hypotension or other evidence of hypoperfusion, consider NOREPINEPHRINE 0.1-2 mcg/kg/min titrated to achieve age appropriate BP.
- If **adrenal insufficiency** is suspected, management per the _Adrenal Insufficiency Protocol._

PEARLS:
- Shock is state in which inadequate tissue perfusion impairs cellular metabolism.
- Shock may be present with a normal blood pressure.
- Tachycardia and tachypnea are early signs of shock in children.
- Children maintain BP and cardiac output by increasing HR and vasoconstriction, even with the loss of significant volume.
- Heart rate is affected by pain, anxiety, medications, and hydration status.
- Heart rate may not be elevated in septic hypothermic patients.
- Septic shock has a 50% mortality rate and must be treated aggressively. Early goal directed therapy consisting of IV fluid administration and early antibiotics reduces mortality in septic patients.
- Common sites/sources for infection include the urinary tract, lungs, skin, GI tract and indwelling catheters and devices.
- Systemic inflammatory response syndrome (SIRS) is a clinical syndrome that results from a deregulated inflammatory response or to a noninfectious insult.
- EtCO2 correlates to serum lactate levels. An EtCO2 of ≤25 mmHg is strongly correlated to a serum lactate of 4.0 mm/L.
- Common sites/sources for infection include the urinary tract, lungs, skin, GI tract and indwelling catheters and devices.
- **Cold shock** = CRT >3 sec, diminished peripheral pulses, mottled or cool extremities.
- **Warm shock** = CRT <1 sec, bounding peripheral pulses, wide pulse pressure, warm/ flushed extremities.
- Patients with sepsis are at an increased risk for acute lung injury (ALI) related to positive pressure ventilation. If positive pressure ventilation is required, avoid excessive tidal volumes.
- Patients at increased risk for adrenal insufficiency (AI) include those with HIV/AIDS, a history of chronic steroid use (current, recently discontinued or remote), Addison disease, or dehydration.
- In smaller patients (<10 -20 kg) rapid boluses may be most easily administer is a 60 ml syringe in utilizing the push/pull method.

Recognition:
- Sickle cell crisis occurs in patients with sickle cell disease (a family of diseases including a hemolytic anemia known as sickle cell anemia).
- Acute pain (often described as sharp, intense, stabbing, or throbbing) in any part of the body, but most commonly in the joints, lower back, legs, arms, abdomen, or chest.
- Patients may also present with stroke, priapism, acute chest syndrome (fever, chest pain, hypoxemia, respiratory symptoms) skin ulcerations, or ocular manifestations.
- In young children, additional presentations include dactylitis (swollen digits), splenic sequestration (enlarged spleen, pallor, hypotension) and aplastic crisis (bone marrow shutdown, noted by pallor)

E

- Routine patient care.
- Consider the administration of supplemental OXYGEN.
- Treat patients with suspected stroke per *Ischemic Stroke Protocol.*
- Transport patient to the nearest appropriate Hospital Emergency Facility, consider transporting the pediatric patient to a Pediatric Specialty Care Hospital.

A
C
P

- Consider analgesia per the age appropriate *Patient Comfort Protocol.*
- NORMAL SALINE 20 ml/kg IV if the patient appears dehydrated.

PEARLS:
- Most of the manifestations of sickle cell crisis are secondary to vascular occlusion and inflammation.
- Sickle cell crisis may be precipitated by a number of physical and environmental factors. Treat any identified possible precipitating factors.
- Pulse oximetry may be unreliable in patients with sickle cell crisis.
- Patients with sickle cell disease are at high risk for life threatening disorders at a very young age.

Recognition:
- Transient, self-limited loss of consciousness with an inability to maintain postural tone that is followed by spontaneous recovery.

E

- Routine patient care.
- Consider *Spinal Motion Restriction Precautions* if indicated.
- Perform blood glucose (bG) analysis, treat as indicated following age appropriate *Diabetic Emergencies Protocol*.
- Assess for traumatic injuries if associated with a fall.
- If patient is hypotensive, manage per the *General Shock and Hypotension Protocol*
- Transport the patient to the nearest appropriate Hospital Emergency Facility.

A C P

- Acquire a multi-lead ECG, manage per *Chest Pain - Acute Coronary System - STEMI Protocol* or the appropriate *Cardiac Dysrhythmia Protocol(s)* as indicated.
- Institute continuous ECG monitoring.

PEARLS:
- History and physical examination are the most specific and sensitive ways of evaluating syncope.
- Patients with any of the following require a thorough in hospital work-up: exertional onset, chest pain, dyspnea, back pain, palpitations, severe headache, focal neurologic deficits, diplopia, ataxia, or dysarthria.
- Consider all syncope to be of cardiac origin until proven otherwise.
- While often thought as benign, syncope can be the sign of more serious medical emergency.
- Syncope that occurs during exercise often indicates an ominous cardiac cause. Patients should be evaluated at the ED. Syncope that occurs following exercise is almost always vasovagal and benign.
- Prolonged QTc (generally >500ms) and Brugada Syndrome (incomplete RBBB pattern in V1/ V2 with ST segment elevation) should be considered in all patients. Consider vasovagal response, GI bleed, dysrhythmia, ectopic pregnancy and aortic aneurysm or dissection as possible causes of syncope.
- Greater than 25% of geriatric syncope is cardiac dysrhythmia related.

2.25 Excited Delirium Syndrome (ExDS)

Recognition:
- Individual ≥16 yo exhibiting bizarre and aggressive behavior including agitation and ≥6 of the following; increased tolerance to pain, tachypnea, diaphoresis, agitation, warm/hot skin to touch, non-compliance to police presence/commands, absence of fatigue, unusual strength, naked or dressed inappropriately for conditions, unusual attraction to glass or mirrors, or keening (unintelligible animal like noises). ExDS is often associated with the chronic use of sympathomimetic drugs (cocaine, methamphetamine, PCP).

E

- Consider safety of EMS providers first. If law enforcement is not present at the scene, consider staging in a safe area until the arrival of law enforcement.
- Routine patient care.
- Control and/or restrain patient as soon as possible to reduce the risks of prolonged struggle. Physical restraint should be performed/assisted by law enforcement when available. If restraints are applied by EMS providers, follow the *Patient Restraint Procedure Protocol.*
- Remove/limit unnecessary stimuli where possible, including warning lights/sirens.
- Manage per appropriate protocols as indicated (*Altered Mental Status*, *Patient in Police Custody*).
- Transport the patient to the nearest appropriate Hospital Emergency Facility.

A C

- MIDAZOLAM 2.5-5 mg IV [5 mg IM/IN] (repeat as needed if the is SBP >100 to a max of 10 mg).
- Initiate continuous monitoring of EtCO2, SpO_2 and ECG.
- NORMAL SALINE 1000 ml IV (may repeat X1).
- Initiate cooling measures if temperature >101˚.
- In the event of cardiac arrest, SODIUM BICARBONATE 50 mEq IV and manage per the age appropriate *Cardiac Arrest Protocol.*

P

- KETAMINE **(preferred)** 4 mg/kg IM (max 400 mg) **or** 2 mg/kg IV (max 200 mg) **or** HALOPERIDOL 5-10 mg IM/IV (may repeat X1) **or** MIDAZOLAM 2.5-5 mg IV [5 mg IM/IN] (may repeat if SBP >100 to max of 10 mg).
- Initiate continuous monitoring of $EtCO_2$, SpO_2 and ECG.
- If sedation is required subsequent to KETAMINE administration, MIDAZOLAM 2.5 mg IV.
- For hypersalivation following KETAMINE administration, ATROPINE SULFATE 0.5 mg IV//IM.
- NORMAL SALINE 1000 ml IV (may repeat X1).
- Initiate cooling measures if temperature >101˚.
- In the event of cardiac arrest, SODIUM BICARBONATE 50 mEq IV and manage per the age appropriate *Cardiac Arrest Protocol.*

PEARLS:

- Be sure to consider all possible medical/trauma etiologies for behavior (hypoglycemia, toxicological, hypoxic, head injury).
- Risk factors associated with excited delirium syndrome include male gender (average age 36), stimulant drug abuse (cocaine and to a lesser extent methamphetamine, PCP, and LSD), history of chronic stimulant drug abuse with recent acute bridge, and preexisting psychiatric disorder (schizophrenia, bipolar disorder).
- Do not position or transport any restrained patient in such a way (e.g. prone) that could negatively affect the patient's respiratory or circulatory status.
- Any patient who is handcuffed or restrained by law enforcement and transported by EMS must be accompanied by a law enforcement officer.
- **For IM administration, a 100 mg/ml concentration of Ketamine is preferred.**
- **Ketamine in a concentration of 100 mg/ml must be diluted 1:1 with 0.9% saline, D5W or sterile water creating a 50 mg/ml concentration prior to IV use.**
- **When administered IV, Ketamine should be administered slowly (over a period of 60 seconds). More rapid administration may result in respiratory depression and enhanced pressor response.**
- **While uncommon with doses ≤4 mg/kg, laryngospasm may occur following the administration of ketamine, it is usually easily managed with positive pressure ventilation.**
- Continuous monitoring of $EtCO_2$ (waveform) and SpO_2 are mandatory in patients who have received midazolam or ketamine for sedation. When clinically feasible, ECG monitoring is required for patients that have received droperidol or haloperidol.
- It may be necessary/appropriate to administer IM injections through clothing in extremely agitated patients.
- Patients with ExDS are at risk for acidosis, rhabdomyolysis and hyperkalemia.

Section 3: Cardiac Protocols

Recognition:
- Respiratory distress, dyspnea on exertion, orthopnea, bilateral crackles on lung auscultation, jugular venous distention, peripheral edema, diaphoresis, hypotension, shock, chest pain/discomfort.

E
- Routine patient care.
- Place patient in upright position as tolerated and as BP allows.
- For patients with **respiratory distress, crackles on lung auscultation, or SPO$_2$ less than 92%**, if the SBP ≥ 90, provide continuous positive airway pressure (CPAP) up to 10 cmH$_2$0 as tolerated by the patient.
- ASPIRIN 81 mg X 4 orally (chewed), unless allergic or unable to swallow safely.
- If the patient has **chest pain or discomfort** manage per the _Chest Pain- Acute Coronary Syndrome-STEMI Protocol._
- If the patient is **hypotensive or has signs of cardiogenic shock or poor perfusion**, manage per the age appropriate _General Shock and Hypotension Protocol._
- Transport the patient to the nearest appropriate Hospital Emergency Facility, consider ALS intercept if available.

A
C
- NITROGLYCERIN 0.4 mg SL (tablet, lingual spray/powder) every 5 min if the SBP is >100.
- If the transport time is ≥ 30 min **and** the patient takes oral furosemide **and** the patient is normotensive (SBP ≥100), consider administering the patients daily dose of FUROSEMIDE (max 80 mg) IV.
- If necessary for CPAP mask compliance, contact **MEDICAL CONTROL** for authorization to administer MIDAZOLAM 1-2 mg IV.

P
- NITROGLYCERIN 0.4 mg SL (tablet or lingual spray/powder) every 5 min if the SBP is >100.
- If the patient is hypertensive or in severe distress, NITROGLYCERIN IV infusion starting at 100 mcg/min and titrated rapidly to symptom improvement or hemodynamics (30% reduction in MAP). Discontinue infusion if the SBP is <100.
- ENALAPRILAT 1.25 mg IV for the patient unresponsive to nitroglycerin with a SBP >140.
- MIDAZOLAM 1-2 mg IV if needed to enhance CPAP compliance.
- If the transport time is ≥ 30 min and the patient takes oral furosemide and the patient is normotensive (SBP ≥100), consider administering the patients daily dose of FUROSEMIDE (max 80 mg) IV.

PEARLS:

- The primary management of ADHF is focused on reducing cardiac afterload, increasing renal perfusion and cardiac output. This is accomplished in the filed with the early application of CPAP and the administration of vasodilators (NTG, ACEI).
- Diuretics (furosemide) and opioids (morphine sulfate) have not been shown to improve outcomes in the EMS management of patients with pulmonary edema. Furosemide should be considered a 2nd tier intervention and reserved for extenuating circumstances where transportation may be delayed.
- NTG should not be administered to patients who have used sildenafil (Viagra, Revatio) or vardenafil (Levitra) in the past 24 hours or tadalafil (Cialis, Adcirca) in the past 36 hours. Revatio is prescribed for pulmonary hypertension.
- Consider acute coronary syndrome in all patients with ADHF/pulmonary edema. Manage per the Chest Pain - Acute Coronary Syndrome - STEMI Protocol as indicated.
- If ADHF/PE is resulting from inferior wall ischemia/infarction, consider obtaining a right sided ECG to identify right ventricular (RV) infarction. NTG should be used cautiously, if at all in patients with RV infarction. f hypotension develops following the administration of NTG, the administration of an IV fluid bolus may be necessary.
- The administration of benzodiazepines to patients requiring CPAP may result in further respiratory depression, particularly in those with a history of recent drug or alcohol ingestion. All efforts at verbal coaching should be utilized to enhance CPAP compliance prior administering benzodiazepines.
- Transdermal administration of NTG (Nitropaste) has a slow onset of action and erratic absorption.
- One dose of SL NTG is equivalent to 60-80 mcg/minute.

Recognition:
- Patient with a complaint of chest pain/discomfort consistent with a cardiac etiology or other known or suspected anginal equivalent.
- STEMI: ST elevation in ≥2 contiguous leads of ≥ 2mm in males or ≥ 1.5mm in females in leads V2-V3 and/or of ≥1 mm in other contiguous chest leads or the limb leads
- Posterior MI: ST depression >1mm in V1-V3 with a dominant R wave (R/S ratio >1) in V2 and upright T waves.
- New onset left bundle branch block (must be evaluated in context).

E
- Routine patient care.
- ASPIRIN 81 mg X 4 orally (chewed), unless allergic or unable to swallow safely.
- For patients <u>prescribed</u> NITROGLYCERIN with a SBP ≥ 100 mmHg, administer 0.3/0.4 mg SL (tabs or lingual spray/powder) of their own medication, may repeat every 5 min to a max of 3 doses if the SBP remains ≥ 100.
- If equipment resources are available, acquire a multi-lead (≥ 12 lead) ECG and transmit ECG to **MEDICAL CONTROL** at the <u>nearest PCI capable facility</u> for assistance with interpretation (see Table 2 Point of Entry - Specialized Hospital Emergency Facilities in Routine Patient Care).
- Transport the patient to the nearest appropriate Hospital Emergency Facility.

A
C
- Acquire a multi-lead (≥ 12 lead) ECG and transmit ECG to **MEDICAL CONTROL** at the <u>nearest PCI capable facility</u> for assistance with interpretation (see Table 2 Point of Entry - Specialized Hospital Emergency Facilities in Routine Patient Care).
- If ECG is suggestive of STEMI:
  - Limit on scene time to ≤ 10 min.
  - Triage patient to the nearest PCI capable facility.
  - Provide immediate notification/**CODE STEMI** to the receiving facility (to expedite registration, provide the patient's name, DOB and if available, patient ID number).
- If the initial ECG is not diagnostic but suspicion is high for STEMI, obtain serial ECGs at 5-10 min intervals and retransmit if changes are noted.
- Consider NITROGLYCERIN 0.4 mg SL (tablet or lingual spray/powder) every 5 min if the SBP is >100.
- Analgesia as indicated per the age appropriate *Patient Comfort Protocol*.
- If the patient is hypotensive or has signs of cardiogenic shock/poor perfusion, manage per the age appropriate *General Shock and Hypotension Protocol*
- Manage dysrhythmias per the age appropriate *Cardiac Dysrhythmia Protocol(s).*

P
- Acquire a multi-lead (≥ 12 lead) ECG; if the ECG is suggestive of STEMI:
  - Limit on scene time to ≤ 10 min.
  - Triage patient to the nearest PCI capable facility (see Table 2 Point of Entry Specialized Hospital Emergency Facilities in Routine Patient Care).
  - Consider consulting **MEDICAL CONTROL** if assistance with ECG interpretation is needed.
  - Provide immediate notification/**CODE STEMI** to the receiving facility (to expedite registration, provide the patient's name, DOB and if available, patient ID number).
- If initial ECG is not diagnostic but suspicion for STEMI is high, obtain serial ECGs at 5-10 min intervals.
- Consider NITROGLYCERIN 0.4 mg SL (tablet or lingual spray/powder) every 5 min if the SBP is >100.
- Consider NITROGYLCERIN by IV infusion starting at 5 to 10 µg/min and increase while titrating to effect.
- Analgesia as indicated per the age appropriate *Patient Comfort Protocol*.
- If patient is hypotensive or has signs of cardiogenic shock/poor perfusion, also manage per the age appropriate *General Shock and Hypotension Protocol*.
- Manage dysrhythmias per the age appropriate *Cardiac Dysrhythmia Protocol(s).*

PEARLS:

- Oral agents and IV Ketorolac are not indicated for the management of chest pain/discomfort of suspected cardiac etiology.
- Patients without STEMI should be transported to the nearest appropriate Hospital Facility.
- A copy of all acquired multi-lead ECGs must be provided to the receiving facility as part of the prehospital medical record. Additionally, a copy must be maintained as part of the EMS medical record by the licensed ambulance service.
- Providers should maintain a low threshold for acquiring a multi-lead ECG in elderly, female or diabetic patients with vague/non-specific or upper GI (nausea, GI distress etc.) complaints. A low threshold also applies to patients with a history of coronary artery disease, HTN, smoking and other cardiac risk factors.
- Reperfusion is time-critical for STEMI patients, with a linear relationship between time to reperfusion and mortality.
- Posterior infarction accompanies 15-20% of STEMIs, usually occurring in the context of an inferior or lateral infarction.
- While not meeting the criteria for STEMI, STE in lead aVR with global STD is concerning for a high risk proximal LAD or left main lesion.
- Right ventricular infarction (RVI) complicates up to 40% of inferior wall STEMI. Consider performing a right sided ECG in patients with ECG findings suggestive of inferior wall STEMI.
- The following are suggestive of RVI: STE in V1, STE in lead III > lead II, STE in V4R.
- STE in V4R has a sensitivity of 88%, specificity of 78% and diagnostic accuracy of 83% in the diagnosis of RV MI.
- Patients with RVI are very preload sensitive and can develop severe hypotension in response to nitrates or other preload-reducing agents. If hypotension develops following the administration of NTG, treat hypotension with an IV fluid bolus. NTG should not be used in patients with RV infarction.
- NTG should not be administered to patients who have used sildenafil (Viagra, Revatio) or vardenafil (Levitra) in the past 24 hours or tadalafil (Cialis, Adcirca) in the past 36 hours. Revatio is prescribed for pulmonary hypertension.
- Unless there is a known history of coronary disease, NTG should not be administered to patients < 16yo without consultation of medical control.

- Routine patient care.
- In situations where adequate bystander cardiopulmonary resuscitation (CPR) [good quality compressions/other care] is ongoing upon EMS arrival, proceed with BLS or ALS assistance as below. If no bystander care is in progress, begin CPR following current AHA ECC Guidelines.
- A defibrillator (AED or manual) should be applied as soon as available and ECG rhythm analysis should immediately follow. If indicated (VF/VT), electrical therapy should be delivered without delay. The initial shock should be delivered at the defibrillator manufacture's recommended energy dose. Subsequent shocks should be administered as indicated every 2 minutes, interposed between two minute CPR duty cycles.
- Continuous compressions and delivery of electrical therapy should take priority over other care.
- Maintain good quality continuous compressions by switching providers every 2 minutes. **Rhythm checks should occur at this time and pauses should be limited to ≤ 5 seconds.**
- Pre-charge the defibrillator at 1:45 sec of each duty cycles to minimize pre-shock pauses if electrical therapy is indicated.
- CPR should be resumed immediately following the delivery of electrical therapy without a pulse check.
- If an automated CPR device (load-distributing or piston) is utilized, the time for application should be minimized.
- Continuous inline waveform capnography may be helpful in determining the quality of chest compressions identifying return of spontaneous circulation (ROSC).
- If the $EtCO_2$ is < 10 mmHg, attempt to improve CPR quality.
- Avoid over-ventilation; ventilation should occur at a rate of 10 bpm.
- Advanced airway management (endotracheal intubation or placement of BIAD) should not result in interruption of chest compressions.
- **Regardless of proximity to a receiving facility, absent concern for provider safety, or traumatic etiology for cardiac arrest, continue resuscitative efforts for a <u>minimum</u> of 30 minutes prior to moving the patient to the ambulance or transporting the patient. BLS providers should request ALS if available.**
- If after 30 minutes of resuscitation at the scene, the patient has organized electrical activity or a shockable rhythm or an $EtCO_2 \geq$ 20 mmHg or signs of life (purposeful motor movement, eye opening) during CPR, consideration should be given to continuing resuscitative efforts at the scene.
- Identify possible treatable etiology of cardiac arrest and manage per appropriate protocol(s) as indicated.

Reversible Causes of Cardiac Arrest

- If return of spontaneous circulation (ROSC) is achieved, manage patient per age

Hypovolemia	Tension pneumothorax
Hypoxia	Tamponade (cardiac)
Hydrogen ion (acidosis)	Toxins
Hypothermia	Thrombosis (pulmonary embolism)
Hypo-hyperkalemia	Thrombosis (coronary)

appropriate *Post Cardiac Arrest Care Protocol.*
- Transport the patient to the nearest appropriate Hospital Emergency Facility. Per the *Post Cardiac Arrest Care Protocol*, patients with hemodynamic instability (MAP <65 or SBP <90, electrical instability (recurrent VF/VT, bradycardia recurring TCP or pharmacologic therapy), or STEMI should be transported to PCI capable facility (see *Routine Patient Care Protocol* - Table 2 - Point of Entry - Specialized Hospital Emergency Facilities).

E

A C

- Consider early interosseous placement (if available) or if difficult IV access is anticipated (access site above the diaphragm is preferred).
- EPINEPHRINE (1:10,000) 1 mg IV every 3-5 min.
- For **ventricular fibrillation (VF) or pulseless ventricular tachycardia (VT):**
  - AMIODARONE 300 mg IV, repeat 150 mg for VF/PVT refractory to the first dose and at least one defibrillation attempt.
  - As alternative to amiodarone or for VF/PVT refractory to amiodarone, LIDOCAINE 100 mg IV, repeat every 10 min X2.
- For **refractory VF/PVT:**
  - Change defibrillator pads and apply 2nd set of pads at a new site.
  - Consider *Double Sequential External Defibrillation Procedure Protocol* if resources allow.
- For **pulseless electrical activity (PEA) arrest and suspected hypovolemia,** LACTATED RINGER'S **or** NORMAL SALINE 500-1000 ml IV (may repeat X1).

P

- Consider early interosseous placement (if available) or if difficult IV access is anticipated (access site above the diaphragm is preferred).
- EPINEPHRINE (1:10,000) 1 mg IV every 3-5 min.
- For **ventricular fibrillation (VF) or pulseless ventricular tachycardia (PVT):**
  - AMIODARONE 300 mg IV, repeat 150 mg for VF/VT refractory to the first dose and at least one defibrillation attempt.
  - As alternative to amiodarone or for VF/VT refractory to amiodarone, LIDOCAINE 100 mg IV, repeat every 10 min X2.
- For **recurrent VF/PVT:**
  - PROCAINAMIDE 1.5 gm IV infused over 15 min.
  - METOPROLOL 5 mg IV over 1 min (may repeat every 5 min X3).
- For **refractory VF/PVT:**
  - Change defibrillator pads and applying a 2nd set of pads to a new site.
  - PROCAINAMIDE 1.5 gm IV infused over 15 min.
  - METOPROLOL 5 mg IV over 1 min (may repeat every 5 min X3).
  - Consider *Double Sequential External Defibrillation Procedure Protocol* if resources allow.
- For **pulseless electrical activity (PEA) arrest:**
  - LACTATED RINGER'S **or** NORMAL SALINE 500-1000 ml IV (may repeat X1).
  - Perform needle thoracostomy for suspected tension pneumothorax.
- For **polymorphic ventricular tachycardia (Torsades de Pointes) or suspected hypomagnesemia,** MAGNESIUM SULFATE 2 gm IV.
- For patients with **CPR induced consciousness, consider:**
  - KETAMINE 0.5-1 mg/kg IV **and** MIDAZOLAM 1 mg IV (may repeat every 5-10 min as required).
- Consider placement of a gastric tube to address gastric distention.

PEARLS:

- The focus of resuscitative efforts should be centered on high quality and continuous chest compressions (rate, depth, recoil) with limited interruptions. "Hands on chest time" should be maximized. Peri-shock pauses should minimized. CPR should not be interrupted for endotracheal intubation or placement of a BIAD.
- Absent a traumatic etiology or concerns for EMS provider safety, continue resuscitative efforts at the scene of the cardiac arrest for a period of not less than 30 minutes. Unless for environmental conditions, the <u>patient should not be moved to the ambulance during this time</u> (a definitive airway is not required as long as oxygen delivery and ventilation are achievable utilizing a bag-valve-mask device).
- Attention should be paid to the ventilation rate. Do not hyperventilate, ventilation should occur at a rate of 10 bpm, EtCO2 should be used to guide ventilation.
- In one study, preshock pauses >20 seconds had a 53% lower chance of survival compared to those with preshock pauses less than 10 sec. For every 5 sec increase in shock pause, the chance of survival decreased by 18%.
- Consider possible treatable causes for cardiac arrest. Utilize relevant protocols in conjunction with this protocol when indicated.
- Naloxone has no utility in cardiac arrest, even if secondary to opioid ingestion/overdose.
- For patients with VF/PVT, antiarrhythmic agents (Amiodarone, lidocaine) should be administered after at least one attempt at defibrillation and after the first dose of epinephrine.
- Recurrent VF/VT is defined as being successfully terminated by conventional electrical therapy, but subsequently returns. It should be treated with conventional electrical therapy and antiarrhythmics.
- Refractory VF/VT is not responsive to conventional electrical therapy. It should be treated with conventional electrical therapy, antiarrhythmics and, if resources allow, DSED. In patients with refractory VF/VT, consider replacing defibrillator pads and changing their location.

References

Goto Y, Funada A. Relationship Between the Duration of Cardiopulmonary Resuscitation and Favorable Neurological Outcomes after Out-of-Hospital Cardiac Arrest: A prospective, nationwide, population-based cohort study. J Am Heart Assoc 2016.

Berg RA, Hemphill R, Abella BS, et al. Part 5. Adult Basic Life Support: 2010 American Heart Association Guidelines for Cardiopulmonary Resuscitation and Emergency Cardiovascular Care. Circulation 2010; 122:S685-705.

Bachman MW, Williams JG, Myers JB, et al. Duration of Prehospital Resuscitation for Adult Out of-Hospital Cardiac Arrest: Neurologically intact survival approaches overall survival despite extended efforts. Prehosp Emerg Care. 2014;18(1):134–135.

Rajan, S, et al. Prolonged Cardiopulmonary Resuscitation and Outcomes after Out-of-Hospital Cardiac Arrest. Resuscitation 2016 Aug 17;105:45-51. Epub 2016 May 17.

http://www.ems1.com/cardiac-arrest/articles/2013670-N-C-EMS-increases-CPR-saves-by-staying-on-scence/

http://www.jems.com/articles/supplements/special-topics/ems-state-science-2014/resuscitation-beyond-25-minute-mark.html

Cardiac Arrest Check List

❏ Resuscitation leader identified (has minimal direct patient contact)

❏ Monitor is visible and a dedicated provider is viewing the rhythm with all leads attached

❏ Monitor is in PADS mode

❏ Metronome confirmed continuous compressions are ongoing at 100-120 compressions per minute

❏ Avoid hyperventilation

❏ Defibrillator charged at 1:45 min of 2 min cycle

❏ Defibrillations occurring at 2 minute intervals for shockable rhythms

❏ O2 cylinder with oxygen in it is attached to BVM

❏ EtCO2 waveform is present and value is being monitored, if $EtCO_2$ < 20 quality of chest compressions are evaluated

❏ IV access obtained (IV or IO)*

❏ Underlying cause has been considered and treated early in arrest

❏ Gastric distention addressed with placement of OGT*

❏ Tension PTX has been considered

❏ Family is receiving care and is at the patient's side

*   Procedures are provider level specific

- Routine patient care.
- In situations where adequate bystander cardiopulmonary resuscitation (CPR) [good quality compressions/other care] is ongoing upon EMS arrival, proceed with BLS or ALS assistance as below. If no bystander care is in progress, begin CPR following current age appropriate AHA ECC Guidelines.
- A defibrillator (AED or manual) should be applied as soon as available and ECG rhythm analysis should immediately follow. If indicated (VF/VT), electrical therapy should be delivered without delay:
- The initial shock should be delivered at 2J/kg.
- The second shock should be delivered at 4J/kg.
- Subsequent shocks should delivered at ≥4J/kg (max 10J/kg or adult dose).
- Shocks should be administered as indicated every 2 minutes, interposed between two minute CPR duty cycles.
- Continuous compressions and delivery of electrical therapy should take priority over other care.
- Maintain good quality continuous compressions by switching providers every 2 minutes. Rhythm checks should occur at this time and pauses should be limited to ≤ 5 seconds.
- Pre-charge the defibrillator at 1:45 sec of each duty cycles to minimize pre-shock pauses if electrical therapy is indicated.
- CPR should be resumed immediately following the delivery of electrical therapy without a pulse check.
- Continuous inline waveform capnography may be helpful in determining the quality of chest compressions identifying return of spontaneous circulation (ROSC). If the $EtCO_2$ is < 10 mmHg, attempt to improve CPR quality.
- Avoid over-ventilation; ventilation should occur at a rate of 12-20 bpm.
- Advanced airway management (endotracheal intubation or placement of a BIAD should not result in interruption of chest compressions.
- **Regardless of proximity to a receiving facility, absent concern for provider safety or a traumatic etiology for cardiac arrest, resuscitative efforts should continue for a minimum of 30 minutes prior to moving the patient to the ambulance or transporting the patient. BLS providers should request ALS if available.**
- Identify possible treatable etiology of cardiac arrest and manage per appropriate protocol(s) as indicated.

E

Reversible Causes of Cardiac Arrest

Hypovolemia	Tension pneumothorax
Hypoxia	Tamponade (cardiac)
Hydrogen ion (acidosis)	Toxins
Hypothermia	Thrombosis (pulmonary embolism)
Hypo-hyperkalemia	Thrombosis (coronary)

- If return of spontaneous circulation (ROSC) is achieved, manage patient per age appropriate _Post Cardiac Arrest Care Protocol._
- Transport the patient to the nearest appropriate Hospital Emergency Facility.

A C

- Consider early interosseous placement (if available) or if difficult IV access is anticipated (access site above the diaphragm is preferred if age appropriate).
- EPINEPHRINE (1:10,000) 0.01 mg (10 mcg/0.1 ml)/kg IV every 3-5 min.
- For **ventricular fibrillation (VF) or pulseless ventricular tachycardia (PVT):** unresponsive to initial electrical therapy and one dose of epinephrine:
  - AMIODARONE 5 mg/kg IV (may repeat X2).
  - As an alternative to amiodarone or for VF/PVT refractory to amiodarone, administer LIDOCAINE 1 mg/kg IV (may repeat X1 in 10 min).
  - For refractory VF/PVT, change defibrillator pads and apply a 2nd set of pads at a new site.
- For **pulseless electrical activity (PEA) arrest**, consider LACTATED RINGER'S **or** NORMAL SALINE 20ml/kg IV (may repeat X1).

P

- Consider early interosseous placement (if available) or if difficult IV access is anticipated (access site above the diaphragm is preferred if age appropriate).
- EPINEPHRINE (1:10,000) 0.01 mg (10 mcg/0.1 ml)/kg IV every 3-5 min.
- For **ventricular fibrillation (VF) or pulseless ventricular tachycardia (PVT)** unresponsive to initial electrical therapy and one dose of epinephrine:
  - AMIODARONE 5 mg/kg IV [300 mg max] (may repeat X2).
  - An alternative to amiodarone or for VF/PVT refractory to amiodarone, LIDOCAINE 1 mg/kg IV [100 mg max] (may repeat X1 in 10 min).
  - For refractory VF/PVT, change defibrillator pads and apply 2nd set of pads at a new site.
- For **pulseless electrical activity (PEA) arrest**:
  - Consider LACTATED RINGER'S **or** NORMAL SALINE 20 ml/kg IV (may repeat X1).
  - Perform needle thoracostomy for suspected tension pneumothorax.
- For **polymorphic ventricular tachycardia (Torsades de Pointes)**, consider MAGNESIUM SULFATE 40 mg/kg IV (may repeat every 5 min, 2 gm max).
- Consider placement of a gastric tube to address gastric distention.

PEARLS:
- Most pediatric cardiac arrests are the focus of resuscitative efforts should be centered on high quality and continuous chest compressions (rate, depth, recoil) with limited interruptions.
- "Hands on chest time" should be maximized.
- Peri-shock pauses should minimized.
- CPR should not be interrupted for endotracheal intubation.
- For patients with VF/PVT, antiarrhythmic agents (Amiodarone, lidocaine) should be administered after at least one attempts at defibrillation and after the first dose of epinephrine.
- The bG should be checked in all critically ill pediatric patients.
- Do not hyperventilate, ventilation should occur at a rate of 12-20. EtCO$_2$ should be used to guide ventilation.
- Consider possible treatable causes for cardiac arrest. Utilize relevant protocols in conjunction with this protocol when indicated.

E

- Routine patient care.
- Identify pulse and continuously palpate for 10 minutes. Absent concern for provider safety or a traumatic etiology for the cardiac arrest, the patient should not be moved during this time.
- Repeat primary assessment including vital signs.
- Continue to address specific differentials associated with original dysrhythmia/etiology of arrest.
- Provide airway management as indicated per the age appropriate *Airway Management Protocol.*
- Continue ventilatory support as indicated; do not hyperventilate.
- Decrease and titrate oxygen concentration to maintain SpO2 94-99%.
- Manage hypotension/shock per the age appropriate *General Shock and Hypotension Protocol*.
- Perform blood glucose analysis; manage per the age appropriate *Diabetic Emergencies Protocol as indicated*.
- Transport patient to the nearest appropriate Hospital Emergency Facility. Patients with hemodynamic instability (SBP <90 or MAP <65 or requiring vasopressors), electrical instability (recurrent VF/VT, complex ectopy or bradycardia requiring TCP or pharmacologic therapy) or STEMI should be transported to a PCI capable facility.

A

C

- Utilize waveform capnography to guide ventilation (maintain normocarbia [EtCO2 35-45 mmHg]).
- Acquire a multi-lead ECG (≥ 12 lead); manage as indicated per the *Chest Pain-Acute Coronary Syndrome-STEMI Protocol.*
- Manage cardiac dysrhythmias per the age appropriate *Cardiac Dysrhythmia Protocol(s).*

P

- Utilize waveform capnography to guide ventilation (maintain normocarbia [EtCO2 35-45 mmHg]).
- Perform multi-lead ECG (≥ 12 lead), manage as indicated per the *Chest Pain-Acute Coronary Syndrome-STEMI Protocol.*
- Manage cardiac dysrhythmias per the age appropriate *Cardiac Dysrhythmia Protocol(s).*
- If either agent was effective, consider maintenance infusions of:
  - AMIODARONE at 1 mg/min **or**
  - LIDOCAINE at 2-4 mg/min.
- Consider placement of a gastric tube to address gastric distention.
- For patients with an advanced airway in place, consider sedation and analgesia as indicated per age appropriate *Patient Comfort Protocol.*

PEARLS:

- During any movement of patient, perform a continuous pulse check and continuous ECG monitoring. DO NOT discontinue ECG monitoring at any time (i.e. transfer from ambulance to ED etc).
- Reassess breath sounds, $EtCO_2$ and ECG rhythm after every patient move.
- Hyperventilation can cause significant hypotension and re-arrest during the post resuscitation phase and therefore should be avoided.
- The initial $EtCO_2$ may be elevated immediately post-resuscitation, but will usually normalize. **Do not hyperventilate the patient to achieve a normal $EtCO_2$ (35-45 mmHg)**.
- Titrate vasopressors as needed to maintain a SBP ≥90 or MAP ≥65. Consider treatable etiologies (pneumothorax, hypovolemia, hyperventilation) for post-resuscitation hypotension.
- Patient with long standing hypertension may require a higher MAP to maintain adequate perfusion.
- Both hypoglycemia and hyperglycemia are deleterious during the post-resuscitation phase. Hypoglycemia should be recognized and treated accordingly and hyperglycemia from indiscriminate glucose administration should be avoided.
- Pre-mix vasopressors in anticipation of hypotension.

Post Arrest - ROSC Checklist

❑ DO NOT MOVE the patient for 10 minutes

❑ Assess $EtCO_2$ (should be > 20 with good waveform, do not try to obtain a "normal" $EtCO_2$ by increasing respiratory rate)

❑ Finger on pulse maintain for 10 minutes

❑ Continuous visualization of cardiac monitor rhythm

❑ Check O_2 supply and SpO_2 and titrate to SpO_2 of 94-99%

❑ Obtain 12 lead ECG

❑ Treat bradycardia (< 60 bpm)

❑ Obtain blood pressure (vasopressor agent(s) as indicated)

❑ Evaluate for post-resuscitative airway placement (e.g, endotracheal tube)

❑ Sedation as required (perform and document neurologic examination prior)

❑ When patient is moved, perform CONTINUOUS PULSE CHECK and continuous monitoring of cardiac rhythm

❑ Mask is available for BVM in case advanced airway fails

❑ Once in ambulance, confirm pulse, breath sounds, SpO_2, $EtCO_2$, and cardiac rhythm

❑ Appropriate personnel for transport

❑ Appropriate point of entry (CCL capable facility for STEMI or patients requiring cardiac pacing, pediatric specialty care facility for pediatric patient)

E

- Routine patient care.
- Identify pulse and continuously palpate for 10 minutes. Absent concern for provider safety or a traumatic etiology for the cardiac arrest, the patient should not be moved during this time.
- Repeat primary assessment including vital signs.
- Continue to address specific differentials associated with original dysrhythmia/etiology of arrest.
- Provide airway management as indicated per age appropriate *Airway Management Protocol.*
- Continue ventilatory support as indicated, do not hyperventilate.
- Decrease and titrate oxygen concentration to maintain SpO2 94-99%.
- Manage hypotension/shock per the age appropriate *General Shock and Hypotension Protocol.*
- Perform blood glucose analysis, manage per the age appropriate *Diabetic Emergencies Protocol* as indicated.
- Transport the patient to the nearest appropriate Hospital Emergency Facility. Consider transportation to a Pediatric Specialty Care Facility.

A C

- Utilize waveform capnography to guide ventilation (maintain normocarbia [EtCO2 35-45 mmHg]).
- Acquire a multi-lead ECG (≥ 12 lead), manage per the *Chest Pain - Acute Coronary Syndrome - STEMI Protocol* if indicated.
- Manage cardiac dysrhythmias per the age appropriate *Cardiac Dysrhythmia Protocol(s).*

P

- Utilize waveform capnography to guide ventilation (maintain normocarbia [EtCO2 35-45 mmHg]).
- Acquire a multi-lead ECG (≥ 12 lead), manage per the *Chest Pain - Acute Coronary Syndrome -STEMI Protocol* if indicated.
- Manage cardiac dysrhythmias per the age appropriate *Cardiac Dysrhythmia Protocol(s).*
- Consider placement of a gastric tube to address gastric distention.
- For patients with an advanced airway in place, consider sedation and analgesia as indicated per the age appropriate *Patient Comfort Protocol.*

PEARLS:

- During any movement of patient, perform a continuous pulse check and continuous ECG monitoring. DO NOT discontinue ECG monitoring at any time (i.e. transfer from ambulance to ED etc).
- Reassess breath sounds, EtCO2 and ECG rhythm after every patient move.
- Hyperventilation can cause significant hypotension and re-arrest during the post resuscitation phase and therefore should be avoided.
- The initial EtCO2 may be elevated immediately post-resuscitation, but will usually normalize. **Do not hyperventilate the patient to achieve a normal EtCO2 (35-45 mmHg).**
- Titrate vasopressors as needed to maintain a SBP ≥ 90 or MAP ≥65.
- Consider treatable etiologies (pneumothorax, hypovolemia, hyperventilation) for post-resuscitation hypotension.
- Check and document bG and temperature in all critically ill pediatric patients.
- Patient with long standing hypertension may require a higher MAP to maintain adequate perfusion.
- Both hypoglycemia and hyperglycemia are deleterious during the post-resuscitation phase. Hypoglycemia should be recognized and treated accordingly and hyperglycemia from indiscriminate glucose administration should be avoided.

Recognition:
- Heart rate < 60 with a pulse and evidence of poor perfusion (hypotension, signs or symptoms of shock, altered mental status, chest pain/discomfort, acute congestive heart failure, or syncope related to bradycardia).

E
- Routine patient care.
- Assess appropriateness of heart rate for clinical situation.
- For patients without symptoms/hemodynamic instability, monitor and reassess as indicated.
- Consider treatable etiologies (hypoxia, beta blocker or calcium channel blocker toxicity, electrolyte imbalance) and exit to appropriate protocol if indicated.
- Transport the patient to the nearest appropriate Hospital Emergency Facility.

A C
- ATROPINE SULFATE 0.5-1.0 mg IV, repeat every 3-5 min to achieve a heart rate > 60 (max dose 3 mg) **or**
- Transcutaneous pacing (TCP). Consider analgesia and sedation per the age appropriate *Patient Comfort Protocol.*
- Consider a NORMAL SALINE 250-500 ml IV, repeat as needed (max 2L).

P
- ATROPINE SULFATE 0.5-1.0 mg IV, repeat every 3-5 min to achieve a heart rate >60 [max dose 3 mg] **or**
- Transcutaneous pacing (TCP). Consider analgesia and sedation per the age appropriate *Patient Comfort Protocol.*
- Consider NORMAL SALINE 250-500 ml IV, repeat as needed (max 2L).
- Consider DOPAMINE HCL 2-10 mcg/kg/min IV **or** EPINEPHRINE 2-10 mcg/min IV for bradycardia refractory to atropine sulfate and TCP.

PEARLS:
- Bradycardia associated with symptoms or hemodynamic instability typically occurs with a heart rates <50. Asymptomatic or minimally symptomatic patients do not necessarily require treatment.
- Atropine sulfate should be used cautiously in the setting of myocardial ischemia/infarction as increased heart rate may worsen ischemia or infarction size.
- Atropine sulfate may be ineffective for treating bradycardia related to atrioventricular block (AVB) occurring below the AV node (type II second-degree block or third-degree [complete] block with wide QRS complex). Immediate TCP may be warranted in these patients. Atropine sulfate is also ineffective in patients who are status post cardiac transplant.
- IV fluids should be considered based on the patient's volume status. Do not administer IV fluids to patients with clinical evidence suggesting heart failure (crackles on lung exam, shortness of breath).

Recognition:
- Heart rate < 60 with a pulse and evidence of poor perfusion (hypotension, signs or symptoms of shock, altered mental status, chest pain/discomfort, acute congestive heart failure, or syncope related to bradycardia).

E

- Routine patient care.
- Assess appropriateness of heart rate for clinical situation. For patients without symptoms/hemodynamic instability, monitor and reassess as indicated.
- If HR is < 60 with poor perfusion despite oxygenation and ventilation, initiate external chest compressions and continued ventilation, reevaluate after 2 minutes. If after 2 minutes bradycardia and signs of hemodynamic compromise persist, verify that support is adequate (airway, oxygen source, ventilation). Continue chest compressions and ventilation if bradycardia with poor perfusion persist.
- Continue to ensure adequate oxygenation and ventilation.
- Consider treatable etiologies (hypoxia, beta blocker or calcium channel blocker toxicity, electrolyte imbalance) and exit to appropriate protocol if indicated.
- Transport the patient to the nearest appropriate Hospital Emergency Facility.

A
C

- Consider NORMAL SALINE 20 ml/kg IV (may repeat X2).
- EPINEPHRINE (1:10,000) 0.01 mg/kg [0.1 ml/kg] IV, repeat every 3-5 min.
- For bradycardia believed to be related to increased vagal tone or primary AV conduction block, ATROPINE SULFATE 0.02 mg/kg IV [minimum dose 0.1 mg and max single dose 0.5 mg] (may repeat X1).
- If epinephrine or atropine are ineffective, Transcutaneous Pacing (TCP). Consider analgesia and sedation per the age appropriate *Patient Comfort Protocol*.

P

- Consider NORMAL SALINE 20 ml/kg IV (may repeat X2).
- EPINEPHRINE (1:10,000) 0.01 mg/kg [0.1 ml/kg] IV, repeat every 3-5 min. If IV access is unavailable, consider EPINEPHRINE (1:1,000) 0.1 mg/kg [0.1 ml/kg] via ETT.
- For bradycardia believed to be related to increased vagal tone or primary AV conduction block, ATROPINE SULFATE 0.02 mg/kg IV [minimum dose 0.1 mg and max single dose 0.5 mg] (may repeat X1). If no IV access is available, consider 0.04-0.06 mg/kg via ETT (may repeat X1).
- If epinephrine or atropine are ineffective, Transcutaneous Pacing (TCP). Consider analgesia and sedation per the age appropriate *Patient Comfort Protocol*.
- Consider DOPAMINE 2-10 mcg/kg/min for bradycardia refractory to atropine and TCP.

PEARLS:
- Bradycardia is commonly a pre-terminal physiologic response to hypoxia in the pediatric patient. Initial management should focus on restoring adequate oxygenation and ventilation.
- If pulseless arrest develops manage as per *Pediatric Cardiac Arrest Protocol.*
- Atropine sulfate may be ineffective for treating bradycardia related to atrioventricular block (AVB) occurring below the AV node (type II second-degree block or third-degree [complete] block with wide QRS complex). Immediate TCP may be warranted in these patients. Atropine sulfate is also ineffective in patients who are status post cardiac transplant.
- Heart Rates of 50—60 are not uncommon in athletic adolescents
- Significant Bradycardia (<40-60) may be seen in patients with eating disorders (marker of severity), in these patients fluid boluses should be avoided.

Recognition:
- Narrow complex (QRS ≤ 0.12 sec) tachycardia with a rate of ≥ 150, patient with a pulse.

E

- Routine patient care.
- Transport the patient to nearest appropriate Hospital Emergency Facility.

A
C

- For minimally symptomatic patients, consider close observation and monitoring.
- For the unstable/pre arrest patient, perform synchronized CARDIOVERSION 100-200 J (biphasic) repeat as needed. Consider pre-shock sedation with MIDAZOLAM 2.5-5 mg IV [5 mg IM/IN] **or** DIAZEPAM 2.5-5 mg IV/IM if the SBP ≥100.
- Vagal maneuvers (valsalva).
- If the rhythm is regular, ADENOSINE 12 mg rapid push IV (may repeat X1).
- Contact **MEDICAL CONTROL** for authorization to administer DILTIAZEM 0.25 mg/kg IV [max dose 20 mg], if the SBP ≥ 100 may repeat X1 at 0.35 mg/kg IV [max dose 25 mg].

P

- For minimally symptomatic patients, consider close observation and monitoring.
- For the unstable/pre arrest patient, perform synchronized CARDIOVERSION 100-200 J (biphasic) repeat as needed. Consider pre-shock sedation with MIDAZOLAM 2.5-5 mg IV [5 mg IM/IN] **or** DIAZEPAM 2.5-5 mg IV/IM if the SBP ≥100 **or** if IV access in unavailable, KETAMINE 2mg/kg IM.
- Vagal maneuvers (Valsalva, CSM).
- If the rhythm is regular, ADENOSINE 12 mg rapid push IV (may repeat X1).
- DILTIAZEM 0.25 mg/kg IV [max dose 20 mg], if the SBP ≥ 100 may repeat X1 at 0.35 mg/kg IV [max dose 25 mg]. Consider a maintenance infusion at 5-15 mg/hour **or** METOPROLOL 2.5-5 mg IV over 2-5 min, repeat every 5 min to max of 15 mg to achieve a ventricular rate of 90-100.

PEARLS:

- For patients with sinus tachycardia (HR ≥ 100 to 220 minus the patient's age), search for and treat the underlying cause (anxiety, fever, pain, dehydration, hypoxia, sepsis).
- Adenosine is not the first line agent for the management of atrial fibrillation, but may considered if the patient has a history of conversion with adenosine or to aid rhythm identification.
- Consider a fluid bolus (NORMAL SALINE 500-1000 ml IV/IO) in patients with a history suggestive of dehydration and no evidence of overt heart failure/pulmonary edema.
- Consider CALCIUM CHLORIDE 1 gm SLOW IV/IO prior to administering DILTIAZEM if the BP is tenuous BP (SBP ~100)
- First line agents for rate control in irregular tachycardias (atrial fibrillation) are calcium channel blockers. As per protocol, Adenosine may be considered to assist with diagnosis or if patient has history of Adenosine conversion, but Adenosine is NOT mandated.
- CSM is contraindicated in patients with a history of TIA/stroke, known carotid atherosclerotic disease, or the presence of a carotid bruit.
- If cardioversion is needed and it is impossible to synchronize the defibrillator, deliver an unsynchronized shock (defibrillation).
- The combined use of IV nodal blocking agents (metoprolol, diltiazem) should be used requires caution and should be avoided whenever possible
- Calcium channel blockers (diltiazem, verapamil) are contraindicated in patients with a known diagnosis of or with ECGs findings consistent with Wolfe Parkinson White (WPW) syndrome, Lown Ganong Levine (LGL) or other pre-excitation syndromes.
- Arrhythmias with suspicion of Wolfe-Parkinson-White (WPW) syndrome should be treated with amiodarone following the dosing regimen in the wide complex tachycardia protocol.
- Adenosine administration should be followed by a 10 ml flush of NS.
- The initial dose for adenosine should be reduced to 6 mg and the repeat dose should be reduced to 12 mg in patients taking dipyridamole and those that are status post cardiac transplant.
- Theophylline and caffeine (methylxanthines) competitively antagonize adenosine's effects; an increased dose of adenosine may be required.
- Adenosine is not indicated in patients with sinus tachycardia, atrial fibrillation or atrial flutter.
- Maximum dose of antiarrhythmic should be given prior to changing to another antiarrhythmic.
- The combined use of IV beta blockers and calcium channel blockers should be avoided.

Recognition:
- Tachycardia with a narrow complex QRS (≤ 0.09 sec), patient with a pulse.

E

- Routine patient care.
- Transport the patient to nearest appropriate Hospital Emergency Facility.

A
C

- For the unstable/pre arrest patient, perform synchronized CARDIOVERSION 1 J/kg, may repeat and increase subsequent energy to 2 J/kg. Consider pre-shock sedation with MIDAZOLAM 0.1 mg/kg [2.5 mg max] IV/IM/IN (may repeat X1) [do not administer if <5kg] **or** FENTANYL 2 mcg/kg [75 mcg max] IV/IM/IN.
- For patients with sinus tachycardia (infants usually HR <220, children HR < 180), consider possible underlying etiologies (fever, pain, volume depletion etc.) and manage per the appropriate protocol(s).
- For minimally symptomatic patients, consider close observation and monitoring.
- For the stable patient, obtain a multi-lead (≥ 12) ECG (repeat if conversion occurs).
- Vagal maneuvers (Valsalva).
- Contact **MEDICAL CONTROL** for authorization to administer ADENOSINE 0.1 mg/kg rapid IV push [max 6 mg], may repeat 0.2 mg/kg rapid IV push [max 12 mg].

P

- For the unstable/pre arrest patient, perform synchronized CARDIOVERSION 1 J/kg, may repeat and increase subsequent energy to 2 J/kg. Consider pre-shock sedation with MIDAZOLAM 0.1 mg/kg [2.5 mg max] IV/IM/IN (may repeat X1) [do not administer if <5kg] **or** FENTANYL 2 mcg/kg [75 mcg max] IV/IM/IN **or** if IV access is unavailable, KETAMINE 2 mg/kg IM.
- For patients with sinus tachycardia (infants usually HR <220, children HR < 180), consider possible underlying etiologies (fever, pain, volume depletion etc.) and manage per the appropriate protocol(s).
- For minimally symptomatic patients, consider close observation and monitoring.
- For the stable patient, obtain a multi-lead (≥ 12) ECG (repeat if conversion occurs).
- Vagal maneuvers (Valsalva).
- ADENOSINE 0.1 mg/kg rapid IV push [max 6 mg], may repeat 0.2 mg/kg rapid IV push [max 12 mg].
- If adenosine is ineffective, consider AMIODARONE 5 mg/kg (max 150 mg) IV over 20 min.

PEARLS:

- For patients with sinus tachycardia (HR ≥ 100 to 220 minus the patient's age), search for and treat the underlying cause (anxiety, fever, pain, dehydration, hypoxia, sepsis).
- Consider a fluid bolus (normal saline 20 ml/kg [10 ml/kg in the neonate] IV) in patients with a history suggestive of dehydration and no evidence of overt heart failure/ pulmonary edema.
- If cardioversion is needed and it is impossible to synchronize the defibrillator, deliver an unsynchronized shock (defibrillation).
- Obtaining a continuous running ECG strip during conversion will help aid in the diagnosis of the type of tachyarrhythmia.
- Utilize pediatric defibrillation/multifunction electrical therapy pads for patients <10 kg.
- Vagal maneuvers for infants and small children consist of applying ice or cold water to the face without occluding the airway for a period of 30-60 seconds. For older children, utilize the Valsalva maneuver or have the patient blow through a narrow straw.
- Paramedics may perform carotid sinus massage in children ≥8 yo.
- Adenosine should be administered via a proximal vein and should be followed by a rapid flush of 5 ml normal saline.
- The dose for adenosine should be reduced in patients taking dipyridomole and those that are status post cardiac transplant.
- Theophylline and caffeine (methylxanthines) competitively antagonize adenosine's effects; an increased dose of adenosine may be required.
- Adenosine is contraindicated in patients with sinus tachycardia, atrial fibrillation or atrial flutter.
- The maximum dose of an antiarrhythmic should be given prior to changing to another antiarrhythmic.

Recognition:
- Tachycardia with a wide complex QRS (≥ 0.12 sec), patient with a pulse.

E
- Routine patient care.
- Transport the patient to the nearest appropriate Hospital Emergency Facility.

A C
- For the unstable/pre arrest patient, perform synchronized CARDIOVERSION 100-200 J (biphasic) [if the rhythm is wide and irregular, perform unsynchronized cardioversion], repeat as needed. Consider pre-shock sedation with MIDAZOLAM 2.5-5 mg IV [5 mg IM/IN] **or** DIAZEPAM 2.5-5 mg IV/IM if the SBP ≥100.
- If the rhythm is regular with monomorphic complexes, consider ADENOSINE 12 mg rapid IV push (may repeat X1).
- Contact **MEDICAL CONTROL** for authorization to administer:
  - AMIODARONE 150 mg IV over 10 min (may repeat X1 in 10 min) **or**
  - LIDOCAINE 1-1.5 mg/kg IV (may repeat X1 in 5 min).

P
- For the unstable/pre arrest patient, perform synchronized CARDIOVERSION 100-200 J (biphasic) [if the rhythm is wide and irregular, perform unsynchronized cardioversion], repeat as indicated. Consider pre-shock sedation with MIDAZOLAM 2.5-5 mg IV [5 mg IM/IN] **or** DIAZEPAM 2.5-5 mg IV/IM if the SBP ≥ **or** if IV access in unavailable, KETAMINE 2mg/kg IM.
- If the rhythm is regular with monomorphic complexes, consider ADENOSINE 12 mg rapid IV push (may repeat X1).
- Consider:
  - AMIODARONE 150 mg IV over 10 min (may repeat X1 in 10 min). If amiodarone is effective in terminating the arrhythmia, consider a maintenance infusion of 1 mg/min **or**
  - PROCAINAMIDE 25-50 mg/min until the arrhythmia is suppressed, hypotension ensues, QRS duration increased by 50%, or a cumulative dose of 17 mg/kg is administered **or**
  - LIDOCAINE 1-1.5 mg/kg IV (may repeat X1 in 5 min). If lidocaine is effective in terminating the arrhythmia, consider a maintenance infusion of 2-4 mg/min.
- For polymorphic ventricular tachycardia/Torsades de Pointes, consider MAGNESIUM SULFATE 1-2 gm IV over 5 min.

PEARLS:

- For patients with sinus tachycardia (HR ≥ 100 to 220 minus the patient's age), search for and treat the underlying cause (anxiety, fever, pain, dehydration, hypoxia, sepsis).
- Consider a fluid bolus (NORMAL SALINE 500-1000 ml IV) in patients with a history suggestive of dehydration and <u>no evidence of overt heart failure/pulmonary edema.</u>
- If cardioversion is needed and it is impossible to synchronize the defibrillator, deliver an unsynchronized shock (defibrillation).
- Calcium channel blockers (diltiazem, verapamil) are contraindicated in patients with a known diagnosis of or with ECGs findings consistent with Wolfe Parkinson White (WPW) syndrome. Arrhythmias with suspicion of Wolfe-Parkinson-White (WPW) syndrome should be treated with amiodarone.
- The initial dose for ADENOSINE should be reduced to 6 mg and the repeat dose should be reduced to 12 mg in patients taking dipyridomole and those that are status post cardiac transplant.
- Maximum dose of antiarrhythmic should be given prior to changing to another antiarrhythmic.
- Theophylline and caffeine (methylxanthines) competitively antagonize adenosine's effects; an increased dose of adenosine may be required.
- Adenosine is contraindicated in patients with sinus tachycardia, atrial fibrillation or atrial flutter.
- The presence of the following factors increase the likelihood of ventricular tachycardia in the patient with a wide complex tachycardia: age > 35 yo, history of ischemic heart disease/MI, history of structural heard disease, CHF, cardiomyopathy or a family history of sudden cardiac death.
- If the rhythm is regular with monomorphic QRS complexes, consider VT or SVT.
- If the rhythm is irregular with monomorphic complexes, consider pre-excitation or atrial fibrillation with aberrancy.

Recognition:
- Tachycardia with a wide complex QRS (≥ 0.09 sec), patient with a pulse.

E
- Routine patient care.
- Transport patient to nearest appropriate Hospital Emergency Facility.

A C
- For the unstable/pre arrest patient, perform synchronized CARDIOVERSION 1 J/kg, may repeat and increase subsequent energy to 2 J/kg. Consider pre-shock sedation with MIDAZOLAM 0.1 mg/kg [2.5 mg max] IV/IM/IN (may repeat X1) [do not administer if <5kg] **or** FENTANYL 2 mcg/kg [75 mcg max] IV/IM/IN.
- For the stable patient, obtain a multi-lead (≥ 12) ECG (repeat if conversion occurs).
- Contact **MEDICAL CONTROL** for authorization to administer:
  - AMIODARONE 5 mg/kg (150 mg max) IV over 20 minutes (may repeat X1 in 10 min) **or**
  - LIDOCAINE 1 mg/kg IV (may repeat every 10 min X2 , max cumulative dose 3 mg/kg).

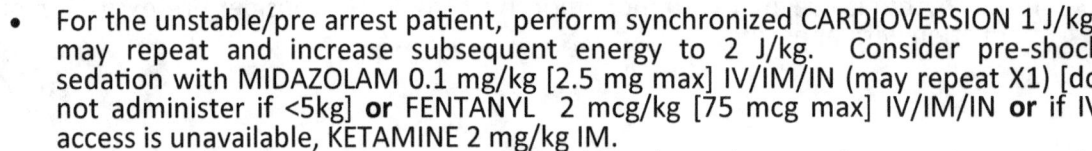

P
- For the unstable/pre arrest patient, perform synchronized CARDIOVERSION 1 J/kg, may repeat and increase subsequent energy to 2 J/kg. Consider pre-shock sedation with MIDAZOLAM 0.1 mg/kg [2.5 mg max] IV/IM/IN (may repeat X1) [do not administer if <5kg] **or** FENTANYL 2 mcg/kg [75 mcg max] IV/IM/IN **or** if IV access is unavailable, KETAMINE 2 mg/kg IM.
- For the stable patient, acquire a multi-lead (≥12) ECG (repeat if conversion occurs).
- Consider:
  - AMIODARONE 5 mg/kg (max dose 150 mg) IV over 20 min (may repeat X1 in 10 min).
  - Initiating transport if refractory to initial therapy.
  - LIDOCAINE 1 mg/kg IV (may repeat every 10 min X2, max cumulative dose 3 mg/kg). If effective, consider an infusion at 20-50 mcg/kg/min.
- For polymorphic ventricular tachycardia/Torsades de Pointes, consider MAGNESIUM SULFATE 50 mg/kg IV (max 2 gm) over 20 min.

PEARLS:
- If cardioversion is needed and it is impossible to synchronize the defibrillator, deliver an unsynchronized shock (defibrillation).
- Utilize pediatric defibrillation/multifunction electrical therapy pads for patients <10 kg.
- For witnessed/monitored VT, try "cough" cardioversion if patient is able to comply.
- Max dose of antiarrhythmic should be given prior to changing antiarrhythmic.
- Place 50 mg/kg magnesium sulfate 50% into 50 ml NS and infuse over 20 minutes.

Recognition:
- Patient presenting with a ventricular assist device.

E

- Routine patient care.
- Determine if you have a patient with a VAD problem, or a patient with a VAD that has a medical problem.
- Assess the patient keeping in minds the following:
  - Skin color and mental status are the best indicators of stability in the VAD patient.
  - A pulse is usually not palpable in the VAD patient. Nearly all VADs are continuous flow devices and is no rhythmic pumping as there is with a functioning ventricle. If the device is a pulsatile flow device, a pulse should be palpable.
  - Blood pressure may or may not be obtainable and auscultated readings are usually unreliable. In a continuous flow device, mean arterial blood pressure (MAP) can be obtained by auscultating with a Doppler to auscultate. The first sound heard during auscultation reflects the MAP. The MAP displayed by an automated non-invasive measurement may also be used. A normal MAP is 60-70 mmHg. If the device is a pulsatile flow device, a blood pressure should be measurable.
  - Data suggest that pulse oximetry readings seem to be accurate, despite the manufacturer stating otherwise.
  - Quantitative waveform capnography should be accurate and can be reflective of pump function (cardiac output). An EtCO2 of < 30 mmHg can be indicative of low perfusion secondary to poor pump function.
  - Temperature should be measured as infection and sepsis are common.
- Assess the VAD:
  - Auscultate over the VAD pump location (this should be just to the left of the epigastrium, immediately below the base of the heart). If the pump is functioning, a low hum should be audible. Do not assume that the pump is functioning just because the control unit looks ok.
  - Palpate the control unit. A hot control unit indicates the pump may be working harder than it should be and often indicates a pump problem such as a thrombosis.
  - Look at the alarms on the control panel. Trouble with the VAD will usually be identified by an alarm. The patient will usually have a resource guide to direct alarm troubleshooting.
  - The patient and family members are generally very knowledgeable about the VAD and troubleshooting problems. Inquire about DNR status. Ask if the device is a continuous or pulsatile flow device. Ask if the patient can receive electrical therapy. Ask if chest compressions can be performed in the event of pump failure.
- If there is **no indication of possible VAD malfunction or failure**, exit to appropriate protocols. Only symptomatic dysrhythmias not at the patient's baseline should be treated. If indicated, place electrical therapy/defibrillation pads away from VAD site and AICD. Call the VAD coordinator and discuss plan with caregivers.
- If the device is a **pulsatile flow device and there is no palpable pulse or detectable blood pressure**, providers should use the device's hand pump to maintain perfusion (family members should be familiar with this).
- If there **is indication of possible device malfunction or failure**, contact the VAD coordinator. Discuss the plan with caregivers.

E

- In the event the **patient is unresponsive, pulseless (no signs of life) with a non-functioning pump** deciding when to initiate chest compressions is very difficult. Chest compressions may cause death by exsanguination if the device becomes dislodged. However, if the pump has stopped, the native heart will not be able to maintain perfusion and the patient will likely die. If a VAD patient is unresponsive and pulseless with a non-functioning VAD and has previously indicated a desire for resuscitative efforts, begin chest compressions and resuscitative efforts. Ensure that all troubleshooting efforts (reconnecting wires, changing batteries, replacing the control unit) have failed prior to starting chest compressions. Contact the VAD Coordinator and MEDICAL CONTROL.

- When **transporting a VAD patient**:
  - Patients without a VAD problem should be transported to the nearest appropriate Hospital Emergency Facility for their condition.
  - Patients with a VAD problem should be transported to their VAD hospital when possible. EMS providers should utilize available resources (private service etc.) to facilitate transportation to the patient's VAD hospital.
  - Always bring the patients resource bag with you. It should contain spare batteries, +/- a spare control unit, contact information for the VAD Coordinator, and directions for equipment and alarm troubleshooting.
  - Always bring spare batteries for the VAD with the patient, even if it is not a VAD related problem. Fresh batteries last 3-5 hours.
  - If the transport is going to be prolonged or it is expected that the patient will be away for a while, try to bring the VAD base power unit with you. Alternately, you can ask the patient's family/caregiver to bring it to the hospital. There may be times when you may need to bring it in the ambulance with the patient and plug it into an inverter to utilize it.

PEARLS:

- Utilize the patient and family as a resource.
- Always contact the VAD Coordinator if there is a VAD related problem or question.
- Common complications in VAD patients include CVA and TIA (incidence up to 25%), bleeding, dysrhythmias, and infection.
- The most common causes of death in VAD patients is sepsis and CVA. Keep these in mind when evaluating a VAD patient with altered mental status.
- VAD patients are preload dependent. Consider that a fluid bolus can often reverse hypoperfusion.

Recommended Resource/Reading:

Mechanical Circulatory Support Organization (MCSO), EMS Guide, January 2015.

Section 4:   Trauma - Environmental - Toxicological Protocols

Recognition:
- Adult patient with blunt or penetrating trauma.

E
- Routine patient care.
- Manage life threatening injuries as they are identified.
- Manage per the following age appropriate protocols as indicated:
  - _Spinal Motion Restriction Precautions_
  - _Airway Management_
  - _External Hemorrhage Control_
  - _Head Trauma -Traumatic Brain Injury_
  - _Extremity and Musculoskeletal Injuries_
  - _Patient Comfort_
  - _Hemorrhagic Shock_
  - _General Shock and Hypotension_
- If open penetrating torso (neck to navel) wounds or open pneumothorax are identified, apply a vented chest seal device (preferred) or an occlusive dressing.
- Impaled objects should be immobilized in place.
- Maintain normothermia (use blankets and increase ambient temperature in the patient compartment of the ambulance).
- On scene time should be limited to ≤ 10 minutes.
- Consider the use of helicopter emergency medical services (HEMS) only if indicated as outlined in the _HEMS Protocol._
- Patients meeting any of the criteria in Table 1 - Trauma Center Triage Criteria (reverse side) should be transported to a Level 1 Trauma Center if the scene to door time is ≤ 45 min and there is no unmanageable airway compromise (patients < 18 yo should be transported to Hasbro Children's Hospital).   If the scene to door time is > 45 min **or** there is unmanageable airway compromise, transport the patient to the nearest Hospital Emergency Facility.

P
- Perform needle thoracostomy for suspected tension pneumothorax.

PEARLS:
- Risk of injury/death increases after age 55.
- A SBP <110 in patients > 65 yo may represent shock.
- Avoid hypothermia as it contributes to coagulopathy.
- Be aware of preexisting medical conditions, especially in the elderly patient.

Table 1 – Level 1 Trauma Center Triage Criteria

Physiologic Criteria
Glosgow coma scale ≤13 Systolic blood pressure ≤ 90 Respiratory rate <10 or >29 bpm or need for ventilatory support
Anatomic Criteria
Penetrating injury to the head, neck, torso or extremities proximal to the elbow or knee Chest wall instability or deformity (flail chest) Two or more proximal long bone fractures Crushed, degloved, mangled, or pulseless extremity Amputation proximal to wrist or ankle Pelvic fracture Open or depressed skull fracture Paralysis
Mechanism of Injury
Fall >20 feet (one story is 10 ft) High risk vehicle crash with intrusion (including roof) >12 inches in occupant compartment or >18 inches any site **or** ejection (partial or complete) from vehicle **or** Death of occupant in same compartment. Auto vs. pedestrian/bicyclist thrown, run over, or with significant (>20 mph) impact Motorcycle crash > 20 mph
Special Patient or System Considerations
Patient taking anticoagulants/antiplatelet agents (excluding ASA) or with a bleeding disorder Age >55 yo Pregnancy >20 weeks EMS provider judgement

Recognition:
- Pediatric patient with blunt or penetrating trauma.

E
- Routine patient care.
- Manage life threatening injuries as they are identified.
- Manage per the following age appropriate protocols as indicated:
  - *Spinal Motion Restriction Precautions*
  - *Airway Management*
  - *External Hemorrhage Control*
  - *Head Trauma -Traumatic Brain Injury*
  - *Extremity and Musculoskeletal Injuries*
  - *Patient Comfort*
  - *Hemorrhagic Shock*
  - *General Shock and Hypotension*
- If open penetrating torso (neck to navel) wounds or open pneumothorax are identified, apply a vented chest seal device (preferred) or an occlusive dressing.
- Impaled objects should be immobilized in place.
- Maintain normothermia (use blankets and increase ambient temperature in the patient compartment of the ambulance).
- On scene time should be limited to ≤ 10 minutes.
- Consider the use of helicopter emergency medical services (HEMS) only if indicated as outlined in the *HEMS Protocol.*
- Patients meeting any of the criteria in Table 1 - Trauma Center Triage Criteria (reverse side) should be transported to a Level 1 Trauma Center if the scene to door time is ≤ 45 min and there is no unmanageable airway compromise (patients < 18 yo should be transported to Hasbro Children's Hospital). If the scene to door time is > 45 min **or** there is unmanageable airway compromise, transport the patient to the nearest Hospital Emergency Facility.

P
- Consider needle thoracostomy for suspected tension pneumothorax.

PEARLS:
- Avoid hypothermia as it contributes to coagulopathy.
- If there is suspected neurologic injury, maintain age appropriate blood pressure.
- Be aware of preexisting medical conditions.

Table 1 – Level 1 Trauma Center Triage Criteria

Physiologic Criteria
Glosgow coma scale ≤13
Systolic blood pressure ≤70
Respiratory rate <20 bpm in an infant <1 yo or need for ventilatory support

Anatomic Criteria
Penetrating injury to the head, neck, torso or extremities proximal to the elbow or knee
Chest wall instability or deformity (flail chest)
Two or more proximal long bone fractures
Crushed, degloved, mangled, or pulseless extremity
Amputation proximal to wrist or ankle
Pelvic fracture
Open or depressed skull fracture
Paralysis

Mechanism of Injury
Fall >10 ft or 2-3 times the height of the child
High risk vehicle crash with intrusion (including roof) >12 inches in occupant compartment or >18 inches any site **or** ejection (partial or complete) from vehicle **or** death of occupant in same compartment.
Auto vs. pedestrian/bicyclist thrown, run over, or with significant (>20 mph) impact
Motorcycle crash > 20 mph

Special Patient or System Considerations
Patient taking anticoagulants/antiplatelet agents (excluding ASA) or with bleeding disorder
EMS provider judgement

Recognition:
- Patient with a history or clinical evidence of blunt or penetrating head trauma.
- Patient with anticoagulant/antiplatelet with Fall

E
- Routine patient care.
- Obtain baseline neurologic examination (GCS, pupils, motor).
- If the patient has clinical evidence of increased intracranial pressure, manage per the *Acute Neurologic Event with Evidence of Increased ICP Protocol.*
- Provide airway management as indicated per age appropriate *Airway Management Protocol.*
- Ventilate the patient as indicated to maintain the EtCO2 35-45 mmHg (avoid hyperventilation), if capnography is not available, ventilate the patient at a rate of 8-10 bpm.
- If the patient has altered mental status, perform blood glucose analysis and manage per the age appropriate *Diabetic Emergencies Protocol.*
- Transport to the nearest appropriate Hospital Emergency Facility. Patients with evidence of TBI should be transported to a Specialty Hospital.

A C P
- Manage hypotension per the age appropriate *General Shock and Hypotension Protocol.*

PEARLS:
- Patients taking anticoagulant/antiplatelet agents with altered mental status following a fall should be transported to an age appropriate Level I Trauma Center.
- Follow and document the patient's neurologic examination. Convey the patient's best neurologic examination and any episodes of hypoxia or hypotension during patient hand off.
- Hypotension is usually a terminal event in isolated head trauma. If hypotension is present, the etiology should be aggressively identified and treated.
- Serial examination of the patient's mental status should be conducted and documented.
- Attempt to obtain information related to use of antiplatelet or anticoagulants by the patient.
- Brain ischemia is worsened by over aggressive hyperventilation. Patients without evidence of increased ICP should be ventilated to maintain normocapnia (EtCO2 35-45 mmHg).
- Short isolated episodes of hypoxia or hypotension should be avoided as they can cause secondary brain injury.
- Hyperglycemia is associated with worsened neurologic outcome, glucose-containing solutions should be administered only as indicated for the treatment of hypoglycemia.

Recognition:
- Patient experiencing a mechanism of injury with risk for spinal injury.

    - Patients meeting any of the following criteria <u>require spinal motion restriction precautions (SMRP)</u>:
        - Patients aged <3 yo or >65 yo.
        - Altered mental status (includes dementia, preexisting brain injury, developmental delay, psychosis).
        - Suspected intoxication (alcohol or other substance).
        - Insurmountable communication barrier.
        - Presence of a distracting injury (an injury believed to be producing pain sufficient to distract the patient from a second (neck) injury.
        - History of underlying spinal disease, surgery (includes fusion or hardware implant) or malignancy with potential for bone metastasis.
        - Motor vehicle crash >60 mph **or** with rollover **or** ejection.
        - Falls >3ft/5 stairs (fall from 3ft above ground surface).
        - Axial load to head/neck (diving, contact sport collision, heavy weighted object falling on head).
        - Significant injury or MOI above clavicles.
        - Injuries involving motorized recreational vehicles (ATV, snowmobile etc.).
        - Bicycle collision/struck.
    - For patients not meeting any of the above criteria, assess as below:

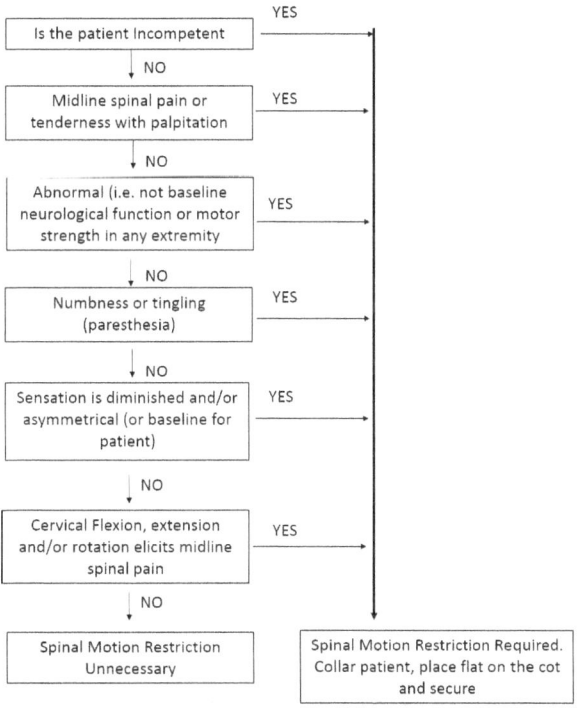

- Apply a rigid cervical collar (for patients who poorly tolerate a rigid cervical collar [e.g. anxiety, shortness of breath], the collar may be replaced with a towel and/or other padding to restrict movement).
- If able, patients in an automobile may "self-extricate" themselves if not prevented by other injuries. Application of rigid cervical collar can be deferred until the patient has self-extricated.
- Ambulatory patients may be allowed to sit and then lie flat on the ambulance stretcher (no "standing takedowns").

E

- Pull sheets, other flexible devices, scoop and scoop like stretchers should be used preferentially to move non-ambulatory patients when appropriate.
- Long spine boards (LSB) should be utilized for extrication and patient movement purposes only. If a LSB is utilized, once the patient is moved onto the ambulance stretcher, the LSB should be removed by using a log roll or lift and slide technique. Patients should only be transported on a LSB if it is necessary for patient safety (e.g. combative patient), if removal from the LSB would delay transportation of an unstable patient or if it is necessary for other treatment priorities (management of increased intracranial pressure)
- Patients should be placed supine on the ambulance stretcher, securely restrained utilizing stretcher straps and instructed to minimize moving their head or neck as much as possible. Elevate the head of the stretcher only if necessary to support respiratory function, patient compliance, or other treatment priorities.
- Consider using SpO2 and EtCO2 to monitor respiratory function.
- Patients with nausea or vomiting may be placed in a lateral recumbent position while maintaining their head in a neutral position using manual stabilization, padding/pillows or the patient's arm. These patients should also be provided antiemetic therapy per the age appropriate *Patient Comfort Protocol*.
- Transfer from ambulance to hospital stretchers and vise-versa should be accomplished while continuing to limit motion of the spine. Slide boards or sheet lifts etc. should used.
- Patients with penetrating trauma such as gunshot or stab wounds do not require SMRP unless a neurologic deficit is noted.
- For **pediatric patients** ≤6 yo or ≤27 kg (60 lbs) requiring SMRP:
  - Transport the patient in a pediatric restraint system as described in the *Routine Patient Care Protocol.* Utilize pediatric restraint systems for older/ larger children when appropriate when they fall within the device's recommended range.
  - Apply padding and a cervical collar as tolerated to minimize spinal motion. Do not force application of a cervical collar if it is not tolerated by the patient. Utilize rolled towels should be utilized if a cervical collar is not tolerated.
  - Avoid any movements that provoke increased spinal movement.
  - If the patient requires significant care (e.g. airway management) that cannot be accomplished in the car seat or pediatric restraint system, remove the patient and secure him/her directly to the stretcher.
  - Patients involved in a motor vehicle crash (MVC) may remain in their own child safety seat for transportation, provided all of the following criteria are met:
    1. The seat has a self-contained harness;
    2. The seat is convertible with both front and rear belt paths;
    3. Inspection does not reveal cracks or deformation;
    4. Vehicle in which the seat was installed was capable of being driven from the scene of the crash;
    5. Vehicle door nearest the child safety seat was undamaged;
    6. There was no airbag deployment; and
    7. The provider is able to ensure appropriate assessment of the patient's posterior.
    8. The safety seat is appropriately secured to the ambulance stretcher, airway seat or bench.

PEARLS:
- Examples of distracting injuries include long bone fractures, visceral injuries, joint dislocations, crush injuries, large burns and thoracic trauma.
- In some circumstances, extrication of a patient utilizing traditional spinal immobilization techniques may result in greater spinal movement or may dangerously delay extrication.

Recognition:

- Patient with external hemorrhage

E

- Routine patient care.
- Apply direct pressure/pressure dressing to injury (including use of a mechanical pressure dressing [e.g. IT clamp]).
- If direct pressure is ineffective or impractical:
  - If the wound is amenable to tourniquet placement (e.g. extremity injury), apply a hemostatic tourniquet.
  - If the wound is not amenable to tourniquet placement (e.g. junctional injury), apply hemostatic agent (combat gauze), if available, with direct pressure.
  - For junctional hemorrhage, if available consider use of a junctional tourniquet and/or an expandable, multi-sponged dressing (XSTAT).
- For hemorrhage originating from a dialysis shunt/fistula:
  - Apply firm finger tip pressure to bleeding site.
  - Apply a pressure dressing (avoid bulky dressing, dressing should not compress the entire shunt/fistula for risk of clotting).
  - If direct pressure and dressing are not effective (i.e. significant hemorrhage continues), apply a tourniquet to the affected extremity. The tourniquet should be applied as remotely from the location of the shunt/fistula as possible.
- Manage per the age appropriate *Hemorrhagic Shock Protocol* as indicated.
- Transport the patient to the nearest appropriate Hospital Emergency Facility.

PEARLS:

- Providers should maintain a low threshold to utilize a tourniquet for severe hemorrhage not easily controlled with direct pressure.
- Tourniquets applied for hemostatic control should both stop bleeding and eliminate the distal pulse.
- Remember to reassess for bleeding after tourniquet application during the course of resuscitation.
- Some injuries may require the application of more than one tourniquet.
- Once applied, tourniquets should not be loosened or removed in the field.
- If available, a pneumatic anti-shock garment may be useful for tamponading bleeding associated with junctional wounds or blast injuries.

Recognition:
- Patient with musculoskeletal injury/complaint.

E

- Routine patient care.
- Manage as indicated per the age appropriate protocols:
  - *Wound Care Procedure*
  - *External Hemorrhage Control*
  - *Hemorrhagic Shock*
  - *Patient Comfort*
- Remove rings, bracelets or other constricting items.
- For **amputations,** clean amputated part, wrap in saline soaked sterile dressing, and place in airtight container. If ice is available, place container on ice (there should be no direct contact between tissue and the ice).
- Immobilize injured extremities utilizing a padded board, wire ladder, vacuum, pillow, plaster/fiberglass or other commercially available splint as appropriate for the specific injury.
- Splint application should result in immobilization of the joint above and below a suspected fracture.
- For **isolated mid-shaft femoral fractures**, the use of a traction splint is the preferred technique for immobilization. A traction splint should not be utilized if a pelvic fracture is suspected.
- Distal peripheral neurovascular status should be assessed and documented in all patients with an extremity injury (this includes prior to and after immobilization).
- **Suspected dislocations and angulated fractures** should generally be immobilized in the position found. However, in some instances when transport time may be prolonged, gentle manipulation may be appropriate to restore distal circulation.
- **Suspected pelvic fractures** should be stabilized utilizing a pelvic binder.
- Cold packs may be applied to affected areas (avoid direct contact with skin).
- Transport the patient to the nearest appropriate Hospital Emergency Facility.

P

- For **open fractures, amputations or grossly contaminated wounds,** CEFAZOLIN 2 gm IV over 3-5 min (peds 25 mg/kg [<80kg 1 gm max dose, >80 kg 2 gm max]).

PEARLS:
- Hip dislocations and knee and elbow fractures have a high incidence of vascular compromise.
- Blood loss may be concealed or not apparent with extremity injuries.
- Cefazolin should not be administered to patients with a history of allergy to cefazolin, other cephalosporin antibiotics, or penicillin.

Recognition:
- Patient in cardiopulmonary arrest with presumed traumatic etiology.

E
- Routine patient care.
- Initiate cardiopulmonary resuscitation (CPR) following current AHA ECC Guidelines while implementing *Spinal Motion Restriction Precautions* if indicated.
- Consider a possible medical etiology for arrest if direct mechanism is not clear (e.g. operator of motor vehicle with minor to moderate vehicular damage observed to suddenly "swerve off the road" prior to collision and now found in cardiac arrest). If uncertainty exists regarding medical or traumatic etiology for cardiac arrest, manage per age appropriate medical *Cardiac Arrest Protocol*.
- Control external hemorrhage following the *External Hemorrhage Control Protocol*.
- Manage other potentially life threatening injuries as per appropriate protocols.
- For patients with a suspected pelvic fracture or those with a mechanism with a high risk for pelvic fracture (pedestrian struck), apply a pelvic binder following the *Pelvic Binder Application Procedure Protocol.*
- Rapidly initiate transportation. If the transport to the nearest Level 1 Trauma Center is <15 min, transport to the nearest Level 1 Trauma Center, if the transport time is >15 min, transport to the nearest Hospital Emergency Facility.

AC
- LACTATED RINGER'S 1L IV bolus [20 ml/kg for peds] (may repeat X1).

P
- LACTATED RINGER'S 1L IV bolus [20 ml/kg for peds] (may repeat X1).
- Perform needle thoracostomy for suspected tension pneumothorax or in any patient with chest trauma.

PEARLS:
- Consider a medical etiology if uncertainty exists regarding etiology for cardiac arrest.
- If the use of spinal motion restriction precautions interferes with the performance of high quality CPR, make reasonable efforts to manually limit patient movement.

Recognition:
- Minor Burn: < 5% total body surface area (TBSA) partial (2nd degree) and full thickness (3rd degree) burns, no inhalation injury, not intubated, normotensive, GCS 14 or greater.
- Serious Burn: 5-15% TBSA partial (2nd degree) and full thickness (3rd degree) burns, suspected inhalation injury or requiring advanced airway management for airway stabilization, hypotension or GCS < 13.
- Critical Burn: > 15% TBSA partial (2nd degree) and full thickness (3rd degree) burns, burns with airway compromise, burns with multiple trauma. Burns involving the face, hands, genitalia, perineum or major joints also meet criteria for a critical burn.

E

- Routine patient care.
- Assess Burn/Concomitant Injury Severity and utilize the Rule of Nines [see reverse] to estimate body surface area (TBSA) affected and assess for evidence of traumatic injuries.
- For **minor burns**:
  - Stop the burning process and remove smoldering and non-adherent clothing.
  - Remove rings, bracelets and other constricting items.
  - Apply dry clean sterile sheet or dressing.
  - Manage as per age appropriate *Trauma related protocols* as indicated.
  - Monitor carboxyhemoglobin (SpCO) if available. If **carbon monoxide (CO) or cyanide exposure** is suspected, also manage per age appropriate protocols. If SpCO is unavailable and CO poisoning is suspected, administer 100% oxygen.
  - Transport the patient to the nearest appropriate Hospital Emergency Facility.
- For **serious or critical burns**:
  - Manage as above.
  - Transport the patient to the nearest Accredited Adult Burn Center. If the patient meets clinical criteria for transportation to a Level I Trauma Center, they should be transported to the nearest Level I Adult Trauma Center as appropriate for their clinical condition unless authorized by **MEDICAL CONTROL**.

A C

- For **minor burns**:
  - Analgesia as indicated per the age appropriate *Patient Comfort Protocol.*
  - Establish IV access if indicated.
- For **serious or critical burns**:
  - Establish IV access, consider two IV sites if greater than 15% TBSA is affected.
  - LACTATED RINGER'S at 300 ml/hr. If the patient is hypotensive or there is evidence of shock, administer a 1000ml bolus, repeat as needed if hypotension or evidence of shock persists.

P

- For **minor burns**:
  - Analgesia as indicated per the age appropriate *Patient Comfort Protocol*.
  - IV access if indicated.
- For **serious or critical burns**:
  - IV access (consider two IV sites if greater than 15% TBSA is affected).
  - LACTATED RINGER'S at 300 ml/hr. If the patient is hypotensive or there is evidence of shock, administer a 1000 ml bolus, repeat as needed if hypotension or evidence of shock persists.
  - Consider early intubation for patients with respiratory distress due to evolving airway obstruction secondary airway edema.
  - For burns >20% BSA with a definitive airway (i.e. ETT in place), consider placement of a gastric tube.

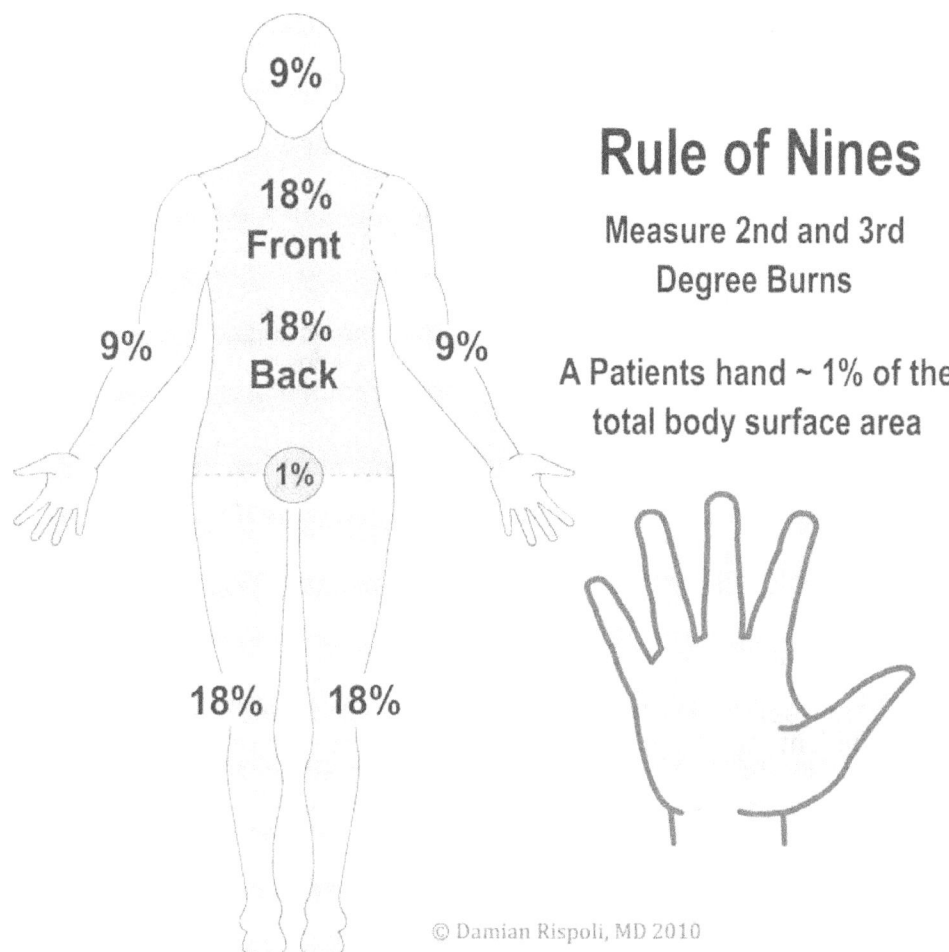

Rule of Nines

Measure 2nd and 3rd
Degree Burns

A Patients hand ~ 1% of the
total body surface area

9%

18%
Front

18%
Back

9%

9%

1%

18%   18%

© Damian Rispoli, MD 2010

*Only areas affected by partial thickness and full thickness burns are used in calculating affected TBSA.

PEARLS:
- Recommended exam: mental status, HEENT, neck, heart, lungs, abdomen, extremities, back, and neurologic.
- Deviation from transporting a serious or critical burn patient to an accredited burn center is permissible if critical interventions such as airway management are indicated, but not possible in the field.
- Patients with thermal burn injuries often have non-burn related traumatic injuries, evaluate for multisystem trauma.
- Burn patients are prone to hypothermia, maintain normothermia.
- Burn patients should not receive intramuscular (IM) injections.

Recognition:

- Minor Burn: < 5% total body surface area (TBSA) partial (2nd degree) and full thickness (3rd degree) burns, no inhalation injury, not intubated, normotensive, GCS 14 or greater.
- Serious Burn: 5-15% TBSA partial (2nd degree) and full thickness (3rd degree) burns, suspected inhalation injury or requiring advanced airway management for airway stabilization, hypotension or GCS < 13.
- Critical Burn: > 15% TBSA partial (2nd degree) and full thickness (3rd degree) burns, burns with airway compromise, burns with multiple trauma. Burns involving the face, hands, genitallia perineum or major joints also meet criteria for a critical burn.

E

- Routine patient care.
- Assess Burn/Concomitant Injury Severity and utilize the pediatric Rule of Nines [see reverse] to estimate body surface area (TBSA) affected and assess for evidence of traumatic injuries.
- For **minor burns:**
  - Stop the burning process and remove smoldering and non-adherent clothing.
  - Remove rings, bracelets and other constricting items.
  - Stop the burning process and remove smoldering and non-adherent clothing.
  - Apply dry clean sterile sheet or dressing.
  - Manage as per age appropriate Trauma related protocols as indicated.
  - Monitor carboxyhemoglobin (SpCO) if available. If **carbon monoxide or cyanide exposure** is suspected, also manage per age appropriate protocols. If SpCO is unavailable and carbon monoxide poisoning is suspected administer 100% oxygen
  - Transport the patient to the nearest appropriate Hospital Facility Emergency Department.
- For **serious or critical burns:**
  - Manage as above.
  - Transport the patient to the nearest Accredited Pediatric Burn Center. If the patient meets clinical criteria for transportation to a Level I Trauma Center, they should be transported to the nearest Level I Pediatric Trauma Center as appropriate for their clinical condition unless authorized by **MEDIC   AL CONTROL**.

A C

- **For minor burns:**
  - Establish IV access if indicated.
  - Analgesia as indicated per the age appropriate *Patient Comfort Protocol*.
- For **serious or critical burns:**
  - Establish IV access, consider two IV sites if greater than 15% TBSA affected.
  - LACTATED RINGER'S at 20 ml/kg/hr. If the patient is hypotensive or there is evidence of shock, administer a 20 ml/kg IVF bolus, repeat as needed if hypotension or evidence of shock persists.
  - Provide analgesia as indicated per the age appropriate *Patient Comfort Protocol*.

P

- For **minor burns:**
  - IV access if indicated.
  - Analgesia as indicated per the age appropriate *Patient Comfort Protocol.*
- For **serious or critical burns:**
  - IV access, consider two IV sites if greater than 15% TBSA affected.
  - LACTATED RINGER'S at 20 ml/kg/hr. If the patient is hypotensive or there is evidence of shock, administer a 20 ml/kg IVF bolus, repeat as needed if hypotension or evidence of shock persists.
  - Early intubation for patients with respiratory distress due to evolving airway obstruction secondary airway edema.
  - For burns >20% BSA with a definitive airway (i.e. ETT in place), consider placement of a gastric tube.

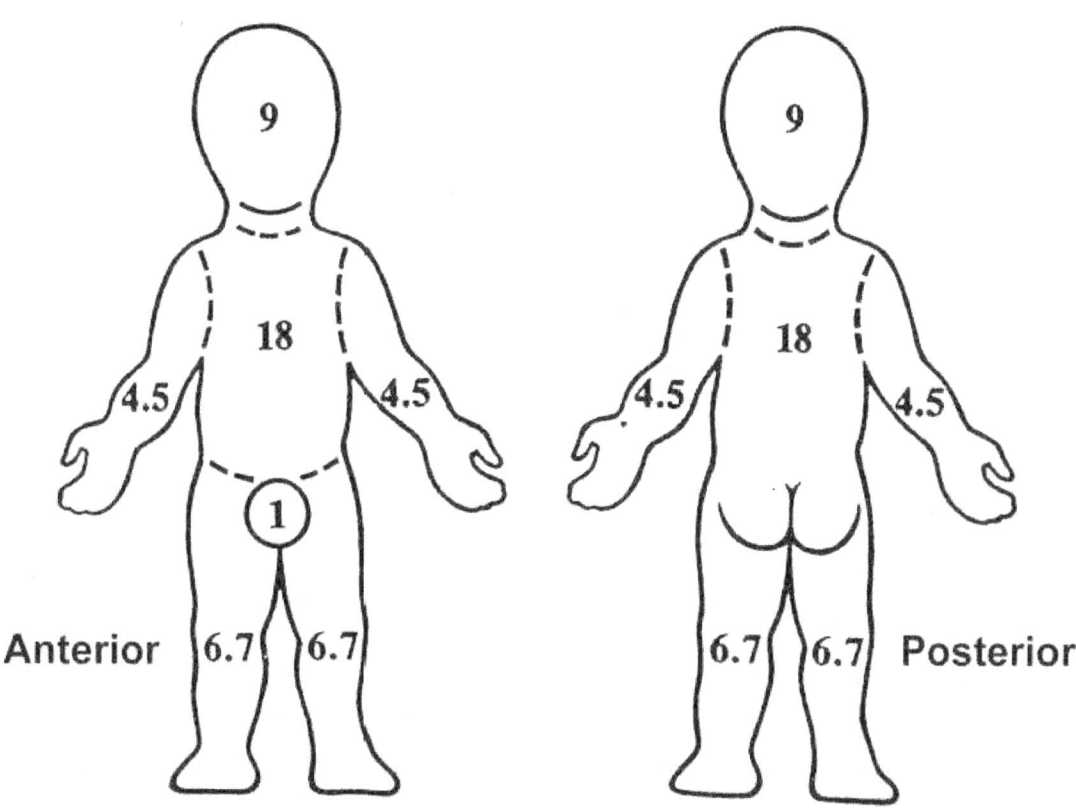

Only areas affected by partial thickness and full thickness burns are used in calculating affected TBSA.

PEARLS:
- Recommended exam: mental status, HEENT, neck, heart, lungs, abdomen, extremities, back, and neurologic.
- Superficial (first degree) burns are characterized by pain and redness, partial thickness (second degree) are characterized by pain and blistering, full thickness (third degree) burns are characterized by painlessness and white/brown coloration.
- Evaluate for the possibility of child abuse in all pediatric patients with burn injuries.
- Deviation from transporting a serious or critical burn patient to an accredited pediatric burn center is permissible if critical interventions such as airway management are indicated, but not possible in the field.
- Patients with thermal burn injuries often have non-burn related traumatic injuries, evaluate for multisystem trauma.
- Burn patients are prone to hypothermia, maintain normothermia.
- Burn patients should not receive intramuscular (IM) injections.

Recognition:
- Partial or full thickness burn injury resulting from contact with a chemical agent or energized electrical source.
- Minor Burn: < 5% total body surface area (TBSA) partial (2nd degree) and full thickness (3rd degree) burns, no inhalation injury, not intubated, normotensive, GCS 14 or greater.
- Serious Burn: 5-15% TBSA partial (2nd degree) and full thickness (3rd degree) burns, suspected inhalation injury or requiring advanced airway management for airway stabilization, hypotension, GCS < 13, circumferential burns of the extremities.
- Critical Burn: > 15% TBSA partial (2nd degree) and full thickness (3rd degree) burns, burns with airway compromise, burns with multiple trauma. Burns involving the face, hands, genitalia, perineum or major joints also meet criteria for a critical burn.

E
- Assure chemical source is not hazardous to EMS providers or other responders. Seek hazardous material resources as indicated.
- Assure electrical source is no longer in contact with the patient prior to touching the patient.
- Routine patient care.
- Assess burn/concomitant injury severity and utilize the Rule of Nines [see age appropriate *Thermal Burn Protocol*] to estimate total body surface area (TBSA) affected and assess for evidence of traumatic injuries.
- Decontaminate **liquid chemical burns** by irrigate affected areas with NORMAL SALINE (clean water or other appropriate solutions or decontaminants may be used as appropriate) for 15 minutes, repeat as needed.
- For exposure to **powdered chemicals**, the affected contact points should be brushed off and then irrigated as above.
- For exposure to **hydrofluoric acid (HF)**, apply (if available) CALCIUM GLUCONATE 2.5 topical gel to all contact points. If the hands or digits are involved, the gel should be placed in an exam glove and the glove worn on the affected hand. If there is ocular exposure, irrigate the affected eye with (if available) CALCIUM GLUCONATE 1% solution.
- Remove non-adherent clothing, rings, bracelets and other constricting items.
- Manage per the age appropriate *Thermal Burn Protocol* and *Trauma* related protocols as indicated.
- Transport the patient to nearest appropriate Hospital Emergency Facility. Patients with **serious or critical burns** require transport nearest age appropriate Accredited Adult Burn Center. If the patient meets clinical criteria for transportation to a Level I Trauma Center, they should be transported to the nearest age appropriate Level I Trauma Center as appropriate for their clinical condition unless authorized by **MEDICAL CONTROL**.

A C
- Initiate cardiac monitoring if indicated, manage per age appropriate *Cardiac Arrest* and *Cardiac Dysrhythmia Protocol(s)* as indicated.

P
- For patients with exposure to **hydrofluoric acid (HF) with** signs and symptoms of hypocalcemia, consider CALCIUM CHLORIDE 1gm slow IV push (may repeat X1). If inhalation injury is suspected, consider (if available) CALCIUM GLUCONATE 2.5% solution via SVN.

PEARLS:
- Recommended exam: mental status, HEENT, neck, heart, lungs, abdomen, extremities, back, neurologic.
- Normal saline or sterile water is preferred for irrigation purposes, however do not delay irrigation if only tap or other water is available. The area should be irrigated as soon as possible with normal saline or sterile water once available.
- Findings suggestive of hypocalcemia include tetany, perioral paraesthesias, increased deep tendon reflexes, laryngospasm, prolonged QT/QTC or polymorphic VT.
- Calcium gluconate gel may be made by mixing 7.5ml of 10% calcium gluconate to 22.5ml of water soluble surgical lubricant.
- For electrical burns, attempt to determine the nature of the electrical source (AC/DC),

Recognition:
- Extremity/body crush, entrapped and crushed under heavy load (structural or trench collapse, heavy equipment/machinery) for >30 minutes.

E
- Routine patient care.
- Request additional resources as needed
- Stage until scene is safe/secure.
- Treat as indicated per age appropriate *Trauma Protocols/Shock Protocol/Hypothermia or Localized Cold Injury Protocol*.
- Remove rings, bracelets or other constricting items.
- Request ALS response if available.
- Provide analgesia per the age appropriate *Patient Comfort Protocol*.
- Transport to the nearest appropriate Hospital Emergency Facility.

A
C
- During extrication, consider D5W (preferred) or NORMAL SALINE 1,000 ml with 150 mEq of SODIUM BICARBONATE added at 1.5 L/hr (peds 20 ml/kg X 3). Once extricated/crush resolved, reduce the infusion rate to 500 ml/hr (peds run at age appropriate maintenance rate).
- An additional IV containing only crystalloid IV fluid should be used if volume resuscitation is required (i.e. hypotension is present) following the age appropriate Shock Protocol.
- For patients with ECG changes suggestive of hyperkalemia or if the patient progresses to cardiac arrest, manage per the age appropriate *Cardiac Arrest Protocol* and contact **MEDICAL CONTROL** for authorization to administer CALCIUM CHLORIDE 1gm IV **or** CALCIUM GLUCONATE 3 gm IV **and** SODIUM BICARBONATE 50 mEq IV.
- Consider ALBUTEROL 5 mg via SVN for patients with suspected hyperkalemia.

P
- During extrication, consider D5W (preferred) **or** NORMAL SALINE 1,000 ml with 150 mEq of SODIUM BICARBONATE added at 1.5 L/hr (peds 20 ml/kg X 3). Once extricated/crush resolved, reduce the infusion rate to 500 ml/hr (peds run at age appropriate maintenance rate).
- An additional IV containing only crystalloid IV fluid should be used if volume resuscitation is required (i.e. hypotension is present) following the age appropriate Shock Protocol.
- For patients with ECG changes suggestive of hyperkalemia **or** if the patient progresses to cardiac arrest, manage per the age appropriate *Cardiac Arrest Protocol* and consider CALCIUM CHLORIDE 1 gm IV **or** CALCIUM GLUCONATE 3 gm IV **and** SODIUM BICARBONATE 50 mEq IV.
- Consider ALBUTEROL 5 mg via SVN for patients with suspected hyperkalemia.

PEARLS:

- Crush injury usually occurs with compression times of 4-6 hrs, but may occur in ≤ 20 min.
- Large volume hydration prior to extrication/removal of crush mechanism is critical to preventing acute kidney injury and mortality. Hydration should be initiated prior to extrication whenever possible
- The larger the crush injury (limbs affected) the greater the likelihood of severe rhabdomyolysis and acute kidney injury.
- Patients with significant crush trauma can have significant fluid shifts (3rd spacing) >12L in the first 48 hours.
- Sodium bicarbonate is preferentially mixed in D5W (150 mEq/1000ml), utilization of 0.9% saline for this purpose is less desirable.
- If possible, monitor patient for signs/symptoms of compartment syndrome (pain, parlor, paresthesia, pulselessness).
- ECG findings suggestive of hyperkalemia include peaked T waves, QRS ≥ 0.12 sec, QT ≥ 0.46 sec or loss of P waves.
- Do not overlook other injuries, airway compromise, hypothermia/hyperthermia. Hypothermia may occur even in settings with warm ambient temperatures.
- Utilize $EtCO_2$ (quantitative waveform capnography) if administering multiple dose of opioids and/or benzodiazepines.
- Confined space treatment should be performed only by appropriately trained personnel.
- If available, document air quality measurement at patient location on patient care report.

Recognition:
- Patient with tooth pain or dental injury.

E
- Routine patient care.
- Evaluate dental or jaw pain for association with cardiac etiology. If cardiac etiology is suspected or cannot be ruled out, obtain multi-lead ECG and exit to age appropriate *Cardiac Protocol(s).*
- If dental injury is associated with significant or multisystem trauma, control bleeding with direct pressure and manage per the age appropriate *Trauma Protocol(s).*
- For **dental related bleeding,** control it with direct pressure by inserting a small gauze dressing rolled into a square and placed into tooth socket with patient closing teeth to exert pressure.
- For **dental avulsions,** place the avulsed tooth in milk, saline, or cell-cultured medium. Do not rub or scrub the avulsed tooth.
- Provide analgesia as indicated per the age appropriate *Patient Comfort Protocol.*
- Transport patient to nearest appropriate Hospital Emergency Facility if indicated.

PEARLS:
- Reimplantation of an avulsed tooth is possible within 4 hours if the tooth is properly cared for.
- Occasionally, pain of cardiac etiology may present as dental or jaw pain.
- All dental associated pain should be associated with a tooth which is tender to tapping, touch or heat/cold sensitivity.

Recognition:
- Patient with ocular complaint or injury.

E
- Routine patient care.
- When feasible, all patients with an ocular related complaint must have their visual acuity (Va) assessed and documented.
- Obtain history of contact lens use, corrective lenses, and safety goggles.
- Contact lenses should be removed when possible.
- For patients with **penetrating ocular injuries,** secure a rigid eye shield over the affected eye. Do not apply gauze type dressings directly over the affected eye as the dressing may adhere to ocular tissue and when it is removed may cause further extravasation of ocular contents. Cover/patch unaffected eye to reduce eye movement. If feasible, elevate the head of bed (HOB) to 30 degrees to decrease intraocular pressure.
- For patients with **ocular impaled objects,** immobilize the object and patch both eyes as above to reduce eye movement (any device used to immobilize the object [cup etc.] should not touch the object). Do not secure a bandage over the top of the cup or other device. If feasible, elevate the head of bed (HOB) to 30 degrees to decrease intraocular pressure.
- For **blunt traumatic injuries,** assess orbital stability and extraocular movements (EOMs). Cover both eyes. If feasible, elevate the HOB to 30 degrees to decrease intraocular pressure.
- For **traumatic enucleation,** cover the open socket with a saline moistened dressing and cover the unaffected eye.
- For **chemical eye injuries,** flush affected eye with a copious amount (minimum 2 L) of water or LACTATED RINGERS or NORMAL SALINE for a period of 15 min (double eversion of the lids will facilitate irrigation). Treat as indicated per the *Chemical and Electrical Burn Injury Protocol.*
- Transport the patient to the nearest appropriate Hospital Emergency Facility.

A
C
- For patients with **penetrating ocular injuries** (including impaled ocular objects), ONDANSETRON 4mg (peds 0.2 mg/kg, max 4 mg) PO/ODT/IV.

P
- For patients with **penetrating ocular injuries** (including impaled ocular objects), ONDANSETRON 4mg [peds 0.2 mg/kg, max 4 mg] PO/ODT/IV.
- For patients with **ocular flash burns related to welding,** consider TETRACAINE 0.5% **or** PROPARACAINE 0.5% 2 drops to the affected eye.
- For patients with **chemical eye injuries:**
  - TETRACAINE 0.5% **or** PROPARACAINE 0.5% 2 drops to the affected eye every 5 min to facilitate irrigation.
  - Consider irrigation utilizing a Morgan Lens®.
  - Utilizing a pH test strip, check the ocular pH five min following irrigation with 2L of fluid. Continue irrigation until the measured pH is 7.0-7.3.

PEARLS:

- Consider consultation with receiving facility regarding timely availability of resources (ophthalmology services). Penetrating global injuries will require emergent in-hospital intervention.
- Patients with severe ocular injuries can present with normal Va.
- Orbital fractures may be associated with cranial nerve injury/entrapment or global injuries. Frequent reassessment of Va and EOMs is warranted in patients with suspected orbital fractures.
- For patients with ocular complaints, determine if there is his a history of tools/equipment with a risk for ocular injury.

Recognition:
- Heat cramps: normal to elevated body temperature, moist skin, weakness, muscle cramps.
- Heat exhaustion: elevated body temperature, cool and moist skin, weakness, anxiety, tachypnea.
- Heat stroke: elevated body temperature [usually >104°], hot and dry skin, <u>altered mental status</u>, +/-hypotension.

E
- Routine patient care.
- Obtain body temperature in all patients with signs/symptoms of hyperthermia or other heat related illness.
- Institute passive cooling measures including removing the patient from the heated environment to a cool environment and remove extra clothing.
- For **heat cramps,** provide oral rehydration as tolerated. Monitor and reassess.
- For **heat exhaustion**, institute active cooling measures including ice applied to the groin, axillae, neck and head, wet sheets placed over the patient, water immersion therapy, and utilization of the mist and fan technique. If the patient is hypotensive, manage per the age appropriate *General Shock and Hypotension Protocol*.
- For **heat stroke**, manage as above for heat exhaustion and per the age appropriate *Airway Management* and *Altered Mental Status Protocols*.
- Transport the patient to nearest appropriate Hospital Emergency Facility.

A C P
- For **heat exhaustion**, NORMAL SALINE 500 ml IV (peds 20 ml/kg), repeat as needed to achieve a SBP >100 or an age appropriate BP in the pediatric patient (adults 2L max, peds 60 ml/kg max).
- For **heat stroke**, NORMAL SALINE (chilled preferred) 1000 ml IV (peds 20ml/kg), repeat as needed to achieve a SBP >100 or an age appropriate BP in the pediatric patient (2L max for adults, 60 ml/kg max for peds).

PEARLS:
- Patients at the extremes of age (young/elderly) are at a higher risk for heat related illness. Other factors predisposing to heat related illness include exposure to increased temperature and/or humidity, poor PO intake, and extreme exertion.
- Certain medications may predispose a patient to heat related illness. These include tricyclic antidepressants, phenothiazines, anticholinergics, selective serotonin uptake inhibitors and alcohol.
- Certain medications may by themselves increase body temperature. These include cocaine, amphetamines and salicylates.
- Heat cramps are not associated with an increased body temperature and are the result of dehydration. They rapidly resolve following rehydration.
- If a commercially available sports drink is used for oral rehydration, it should be diluted 50/50 with water.
- If water immersion therapy is utilized, the patient should have at least their torso submerged, if not their whole body (excluding their head).
- Active cooling measures should be discontinued when the patient's body temperature reaches 101° F to prevent hypothermia.

Recognition:
- Patient with core temperature < 35° C (95°F).
- Patient with cold thermal injury (frostbite).

E

- Routine patient care.
- Provide airway management as indicated per the age appropriate *Airway Management Protocol.*
- In the unresponsive patient, allow 60 seconds for assessment for the presence of a pulse.
- For patients with altered mental status, also manage per the age appropriate *Altered Mental Status Protocol.*
- If the patient is in cardiac arrest or cardiac arrest develops, manage per age appropriate *Cardiac Arrest Protocol*(s) while instituting warming measures below.
- Determine body temperature.
- For patients with a **core temperature ≥ 34-35°C (93.2-95.0°F),** institute external rewarming measures including removing wet clothing , drying the patient, and warm the patient by use of a heat reflective shield or blankets and increasing the ambient temperature.
- For patients with a core **temperature < 34°C/93.2°F,** manage as above and administer (if available) warmed (42-45°C/104-113°F) humidified AIR or OXYGEN.
- For patients with **localized cold thermal injury,** provide general wound care as indicated per the *Wound Care Procedure Protocol.* Do not rub the skin to rewarm and do not allow refreezing to occur. Provide analgesia as indicated per the age appropriate *Patient Comfort Protocol.*
- Transport the patient to the nearest appropriate Hospital Emergency Facility.

A C P

- For patients with a core temperature < 34°C (93.2°F), administer warm (40-42°C/104-107°F) NORMAL SALINE 500 ml IV, repeat as needed unless signs/symptoms of pulmonary edema develop (max 2L).

PEARLS:
- Mild hypothermia = 32-35°C/90-95°F, moderate = 28-32°C/82.4-90°F, severe < 28°C/82.4°F.
- Hypothermia can occur in a warm environment and indoors especially in the elderly, children, patients with impaired thermoregulation (EtOH, burns, sepsis, patients on phenothiazines, barbiturates), patients with endocrine disorders (adrenal insufficiency, hypothyroidism).
- Patients with moderate-severe hypothermia are typically dehydrated, rapid volume expansion with warmed IVF is critical.
- During the course of hypothermia, patients initially manifest tachycardia and then progressively become bradycardic. If paradoxical tachycardia is present, consider hypoglycemia, hypovolemia, and toxic ingestion.
- Waveform capnography may aid in identifying the presence of cardiac output in the severely hypothermic patient without a palpable pulse but with organized electrical activity on the ECG.
- Contrary to popular teaching and belief, a multicenter study (Ann Emerg Med, 1987) found endotracheal intubation does not induce dysrhythmias in the hypothermic patient.
- The myocardium becomes irritable at ≤ 32°, ventricular fibrillation or asystole may occur spontaneously.

Recognition:
- Patient with core temperature < 35° C (95°F).
- Patient with cold thermal injury (frostbite).

E

- Routine patient care.
- Provide airway management as indicated per the age appropriate *Airway Management Protocol.*
- In the unresponsive patient, allow 60 seconds for assessment for the presence of a pulse.
- For patients with altered mental status, also manage per age appropriate *Altered Mental Status Protocol.*
- If the patient is in cardiac arrest or cardiac arrest develops, manage per age appropriate *Cardiac Arrest Protocol*(s) while instituting warming measures below.
- Determine body temperature.
- For patients with a **core temperature ≥ 34-35°C (93.2-95.0°F),** institute external rewarming measures including removing wet clothing , drying the patient, and warm the patient by use of a heat reflective shield or blankets and increasing the ambient temperature.
- For patients with a core **temperature < 34°C/93.2°F,** manage as above and administer (if available) warmed (42-45°C/104-113°F) humidified AIR or OXYGEN.
- For patients with **localized cold thermal injury,** provide general wound care as indicated per the *Wound Care Procedure Protocol.* Do not rub the skin to rewarm and do not allow refreezing to occur. Provide analgesia as indicated per the age appropriate *Patient Comfort Protocol*.
- Transport the patient to the nearest appropriate Hospital Emergency Facility. Consider transportation to a Pediatric Care Specialty Facility.

A C P

- For patients with a core temperature < 34°C (93.2°F), administer warm (40-42°C/104-107°F) NORMAL SALINE 20 ml/kg IV, repeat as needed unless signs/symptoms of pulmonary edema develop (60 ml/kg max).

PEARLS:
- Mild hypothermia = 32-35°C/90-95°F, moderate = 28-32°C/82.4-90°F, severe < 28°C/82.4°F.
- Hypothermia can occur in a warm environment and indoors especially in the elderly, children, patients with impaired thermoregulation (EtOH, burns, sepsis, patients on phenothiazines, barbiturates), patients with endocrine disorders (adrenal insufficiency, hypothyroidism).
- Patients with moderate-severe hypothermia are typically dehydrated, rapid volume expansion with warmed IVF is critical.
- During the course of hypothermia, patients initially manifest tachycardia and then progressively become bradycardic. If paradoxical tachycardia is present, consider hypoglycemia, hypovolemia, and toxic ingestion.
- Waveform capnography may aid in identifying the presence of cardiac output in the severely hypothermic patient without a palpable pulse but with organized electrical activity on the ECG.
- Contrary to popular teaching and belief, a multicenter study (Ann Emerg Med, 1987) found endotracheal intubation does not induce dysrhythmias in the hypothermic patient.
- The myocardium becomes irritable at ≤ 32°, ventricular fibrillation or asystole may occur spontaneously.

Recognition:
- Submersion in water regardless of depth.

E

- Routine patient care.
- Apply *Spinal Motion Restriction Precautions* if indicated.
- For patients in **cardiac arrest**, manage per the age appropriate *Cardiac Arrest Protocol.*
- Manage per the age appropriate *Hypothermia and Local Cold Injury Protocol* as indicated.
- Remove wet clothing, dry and warm the patient.
- Manage per the age appropriate *Airway Management* and *Altered Mental Status Protocol(s) as indicated.*
- If dyspnea or wheezing are present, manage per the age appropriate *Respiratory Distress Protocol* and consider continuous positive airway pressure (CPAP) @ 5-10 cmH20 if tolerated.
- Monitor and reassess the patient.
- Encourage transport and evaluation even if asymptomatic (asymptomatic near drowning patients should be observed for 4-6 hours for the development of complications).
- Transport the patient to the nearest appropriate Hospital Emergency Facility.

PEARLS:
- Cardiac arrest following sudden immersion in near-freezing water may be survivable even after several hours of immersion.
- Hypothermia is commonly associated with submersion/near drowning.
- Post submersion patients who are awake and cooperative with respiratory distress may benefit from continuous positive airway pressure.

Recognition:
- Ear squeeze/sinus squeeze: ear pain, +/- vertigo, +/- recent history of upper respiratory infection (URI), sinus pain, difficulty equilibrating.
- Pulmonary overpressure syndrome (POPS): breath holding on ascent, respiratory distress, subcutaneous emphysema, +/- decreased BS/pneumothorax.
- Arterial gas embolism (AGE): altered mental status, sensory or motor deficits, unequal pupils, vertigo, visual disturbances, and cardiac arrest.
- Decompression sickness (DCS) Type I (non-systemic/musculoskeletal): joint pain (shoulder and elbow most common), back pain, priapism, pruritis, spotted pallor.
- Decompression sickness (DCS) Type II (systemic or neurologic): +/- DCS type I symptoms, paresthesias, paralysis, dizziness/vertigo, dyspnea, hemoptysis, auditory disturbances, chest pain, altered mental status.
- Nitrogen narcosis: "intoxicated appearing", inattention, decreased coordination, poor judgment.

E

- Routine patient care.
- Determine diving history/profile:
  - Time of symptom onset;
  - Parameters of the dive: depth, number of dives, duration, surface intervals;
  - History of air travel following the dive;
  - Rate of ascent;
  - Panic or other factors resulting in rapid ascent;
  - Experience level of the diver;
  - Properly functioning depth gauge;
  - Gas mixture used;
  - Past medical history;
  - Previous episodes of decompression illness; and
  - Use of medications/alcohol.
- Consider consulting the Divers Alert Network (DAN) at 919-684-9111.
- For patients with suspected **"ear or sinus squeeze"**:
  - Provide analgesia per age appropriate *Patient Comfort Protocol*.
  - Patients with ear squeeze require further evaluation to rule out tympanic membrane rupture.
- For patients with suspected **pulmonary over pressure syndrome**:
  - OXYGEN at the highest concentration possible.
  - Treat shock, cardiac arrhythmias or suspected pneumothorax per appropriate protocols.
  - Arrange for emergent recompression therapy (consider utilizing helicopter emergency medical services [HEMS] in consultation with **MEDICAL CONTROL** for transportation of the patient to an emergency hyperbaric oxygen (HBO) capable facility.
- For patients with suspected **arterial gas embolism**:
  - OXYGEN at the highest concentration possible.
  - Arrange for emergent recompression therapy (consider utilizing HEMS in consultation with **MEDICAL CONTROL** for transportation of the patient to an emergency HBO capable facility.
  - Transport the patient in the supine position or on their left side with their head down 30 degrees.
- For patients with suspected **decompression sickness type I**:
  - OXYGEN at the highest concentration possible.
  - Arrange for expedient recompression therapy.
- For patients with suspected **nitrogen narcosis**:
  - OXYGEN at the highest concentration possible.
  - Mange per the age appropriate *Altered Mental Status Protocol* as indicated.

E

- For patients with suspected **decompression sickness type II:**
  - Administer OXYGEN at the highest concentration possible.
  - Arrange for emergent recompression therapy (consider utilizing HEMS in consultation with **MEDICAL CONTROL** for transportation of the patient to an emergency HBO capable facility).

A C P

- For patients with suspected **decompression sickness type II**, administer NORMAL SALINE 1L IV BOLUS, then NORMAL SALINE at 100 ml/hr.
- For patients with suspected **"ear or sinus squeeze":**
  - Provide analgesia per the age appropriate *Patient Comfort Protocol*.
  - OXYMETAZOLINE 2 sprays to each nostril.
  - If available, PSEUDOEPHEDRINE 30 mg PO.
  - Patients with ear squeeze require further evaluation to rule out tympanic membrane rupture.

PEARLS:
- Clinical manifestations of diving related emergencies may be seen during diving or up to 24-48 hours following a dive.
- The main pathologies involved in diving emergencies include barotrauma, decompression illness or the toxic effects of increased partial pressure of gases.
- Any person using SCUBA equipment experiencing neurologic deficits during or after ascent should be suspected of having AGE.
- Signs and symptoms of AGE manifest immediately upon surfacing or during ascent.
- Nitrogen narcosis usually occurs below the surface and is a result of nitrogen's effect on cerebral function. It may be manifested by a diver seemingly taking unnecessary risks during the dive or appearing intoxicated after surfacing.

Recognition:
- Exposure to marine organism with intense localized pain, nausea/vomiting, allergic reaction or anaphylactic reaction.

E
- Routine patient care.
- Manage allergic or anaphylactic reactions per age appropriate *Allergic Reaction-Anaphylaxis Protocol.*
- Identify organism involved.
- Consider consulting the Regional Center for Poison Control and Prevention at 800-222-1222 for advice.
- For patients with **sting ray, lionfish or urchin/starfish related injuries:**
  - Immobilize injury.
  - Remove barb or spine, if large barb in thorax or abdomen, immobilize object.
  - If able, immerse affected area in hot water (110-114°F/43.3 to 45° C).
  - Treat pain per the age appropriate *Patient Comfort Protocol.*
- For patients with **jelly fish or man o' war related injuries:**
  - Immobilize injury.
  - Lift away tentacles (do not rub or brush).
  - If available, apply vinegar rinse, otherwise irrigate with clean seawater. Do not use fresh water or ice.
  - Treat pain per the age appropriate *Patient Comfort Protocol.*
- Transport the patient to the nearest appropriate Hospital Emergency Facility if indicated.

P
- For patients with severe muscle spasms, consider CALCIUM CHLORIDE **or** CALCIUM GLUCONATE 1 gm IV (peds 20 mg/kg) over 3 min.

PEARLS:
- Severe allergic or anaphylactic reactions may occur in seemingly minor envenomations.
- The Lion's mane jellyfish is the most common stinging organism in Rhode Island waters.
- The man o' war is occasionally found in local waters.
- While stingrays and lionfish are both warm water species, both have been identified in local waters (ocean and costal ponds respectively) and could become more frequent visitors as coastal waters warm.
- Urchins and starfish are now less commonly seen local waters and the species of urchins and starfish indigenous to local waters are not venomous. Encounters with toxic urchins and starfish are likely to be related to aquarium inhabitants.

Recognition:
- Bee or wasp sting, spider, snake, feline, canine, or human bite.

E
- Routine patient care.
- Manage allergic or anaphylactic reactions per age appropriate *Allergic Reaction-Anaphylaxis Protocol.*
- Manage patient as indicated per age appropriate per age appropriate *Patient Comfort Protocol.*
- Identify creature/animal involved.
- Consider consulting the **Regional Center for Poison Control and Prevention at 800-222-1222** for advice.
- For **spider bites and bee or wasp stings:**
  - Elevate wound to a neutral position if able.
  - Apply ice/cool packs to affected area.
  - Remove any constricting clothing, bands, or jewelry.
- For **snake bites:**
  - Immobilize extremity if involved.
  - Elevate wound to a neutral position if able.
  - Remove any constricting clothing, bands, or jewelry from affected extremity.
  - Do not apply ice.
  - Mark margin of swelling/redness and time.
- For **feline, canine, or human bites:**
  - Immobilize extremity if involved.
  - Provide wound care per the *Wound Care Procedure Protocol.*
  - Contact and document contact with the jurisdictional animal control officer.
- Transport the patient to the nearest appropriate Hospital Emergency Facility if indicated.

P
- For patients with severe muscle spasms, consider MIDAZOLAM 0.5-2 mg [peds 0.1-0.2 mg/kg] IV over 2-3 min **or** MIDAZOLAM 1-2 mg [peds 0.2 mg/kg] IN **or** MIDAZOLAM 5 mg IM (max 5 mg).

PEARLS:
- Immunocompromised patients (diabetes, chemotherapy, transplant) are at increased risk for infection
- Human bites have a higher infection rate than animal bites due to oral flora.
- Feline bites may progress to infection rapidly due to a specific bacteria (*Pasteurella multicoda*).
- Evidence of infection includes selling, redness, drainage, fever, and red streaking proximal to the wound.
- All carnivore bites carry the risk for Rabies exposure.

Recognition:
- Beta blockers/calcium channel blockers: bradycardia, hypotension, hyperglycemia [CCB].
- Tricyclic antidepressants/sodium channel blocking agents: altered mental status, seizures, hypotension, QRS widening, prolonged QTC, tall R in aVr.
- Opioids: altered mental status, pinpoint pupils, respiratory depression, and hypotension.
- Organosphosphates/nerve agents: salivation, lacrimation, urination, defecation, GI distress, emesis [SLUDGE], bradycardia, bronchospasm, or bronchorrhea.
- Anticholinergics: tachycardia, hyperthermia, dilated pupils, mental status changes.

E
- Routine patient care.
- Obtain history of ingestion/exposure including the substance ingested, route, quantity, time of ingestion, etiology (intentional, accidental, criminal), available medications/toxins in the home and the patient's past medical history and medications.
- For known or suspected **opioid overdose** with respiratory depression/apnea, NALOXONE 2 mg IN, repeat 2.0 mg every 3-5 min until adequate ventilation is restored or 10 mg is administered. Follow the recovery coach algorithm at the end of this protocol.
- For known or suspected exposure to a **nerve or organophosphate agent,** manage per the Nerve Agent-Organophosphate Exposure Protocol.
- Consider contacting the Regional Center for Poison Control and Prevention (800-222-1222) for advice.
- For oral ingestions occurring less than one hour prior to EMS contact, contact **MEDICAL CONTROL** for authorization to administer ACTIVATED CHARCOAL 25 -50 gm PO.
- Transport the patient to the nearest appropriate Hospital Emergency Facility.

A
C
- For suspected **opioid overdose**, if NALOXONE has not been administered or if additive doses are required, NALOXONE 0.4-2 mg IN/IV/IM every 3-5 min until adequate ventilation is restored or 10 mg is administered.
- For suspected **tricyclic antidepressant or other sodium channel blocking agent toxicity,** if seizures are present or the QRS is >0.12 sec, contact **MEDICAL CONTROL** for authorization to administer SODIUM BICARBONATE 50 mEq IV (repeat every 5 min as indicated).
- For suspected **beta blocker or calcium channel blocker (CCB) toxicity**, manage as indicated per _Cardiac Dysrhythmia Protocol(s)_ and contact **MEDICAL CONTROL** for authorization to administer GLUCAGON 1-5 mg IV (may repeat X1 in 15 min) **and** CALCIUM CHLORIDE 1 gm IV **or** CALCIUM GLUCONATE 3 gm IV (may repeat X1).
- For suspected **cyanide toxicity** contact **MEDICAL CONTROL** for authorization to administer (if available) HYDROXOCOBALAMIN 70 mg/kg IV [5 gm max] in 250 ml NS over 15 min (in cardiac arrest the dose may be given over a shorter period).
- For suspected **sympathomimetic/stimulant toxicity**, contact **MEDICAL CONTROL** for authorization to administer MIDAZOLAM 2.5 mg IV/IN [5 mg IM] (may repeat in 5 min, IM dose in 20 min) **or** LORAZEPAM 1 mg IV (may repeat in 5 min).

P

- For suspected **opioid overdose**, if NALOXONE has not been administered or if additive doses are required, NALOXONE 0.4-2 mg IN/IV/IM every 3-5 min until adequate ventilation is restored or 10 mg is administered.
- For suspected **tricyclic antidepressant or other sodium channel blocking agent toxicity:**
  - SODIUM BICARBONATE 50 mEq IV for seizures or if the QRS is > 0.12 sec (may repeat every 5 min).
  - NOREPINEPHRINE 2-20 mcg/min IV for hypotension (SBP < 100).
  - 20% IV FAT EMULSION 1.5 ml/kg IV for profound hemodynamic compromise or cardiac arrest (may repeat X2).
- For suspected **beta blocker or calcium channel blocker (CCB) toxicity:**
  - _Cardiac Dysrhythmia Protocol(s)_ as indicated
  - GLUCAGON 1-5 mg IV (may repeat X1 in 15 min).
  - CALCIUM CHLORIDE 1 gm IV **or** CALCIUM GLUCONATE 3 gm IV (may repeat X1).
  - DOPAMINE 2-20 mcg/kg/min IV for hypotension (SBP <100).
  - 20% FAT EMULSION 1.5 ml/kg IV for profound hemodynamic compromise or cardiac arrest (may repeat X2).
- For suspected **cyanide toxicity:**
  - If available, HYDROXOCOBALAMIN 70 mg/kg IV (5 gm max dose) in 250 ml NS over 15 min (in cardiac arrest the dose may be given over a shorter period).
  - If available, SODIUM THIOSULFATE 12.5 gm/100 ml D5W over 10 min.
- For **sympathomimetic/stimulant toxicity**, MIDAZOLAM 2.5 mg IV/IN [5 mg IM], may repeat in 5 min [IM dose in 20 min] **or** LORAZEPAM 1 mg IV (may repeat in 5 min).
- For **dystonic reactions**, DIPHENHYDRAMINE 25-50 mg IV.
- For any **ingestion of highly lipid soluble medication or substance** with profound hemodynamic compromise or cardiac arrest, 20% FAT EMULSION 1.5 ml/kg IV (may repeat dose X2).

PEARLS:
- For patients with chronic substance abuse, consider contacting a Recovery Coach to meet the patient at the hospital. Recovery coaches may be contacted at 401-415-8833.
- RI General Law 23-10.1-4 allows a police officer to take an individual into protective custody and transport him/her to a hospital if the officer has reason to believe that the individual is intoxicated by drugs other than alcohol and as a result is likely to injure him or herself or others if allowed to be at liberty pending examination by a physician.
- Consider polydrug overdose in all patients.
- Identification of toxidromes may help identify toxins and guide therapy.
- Consider cyanide toxicity in any patient with smoke inhalation with altered mental status and unexplained hypotension.
- Patients with suspected cyanide toxicity should be administered oxygen at the highest concentration available.
- Highly lipid soluble cardiotoxic agents include, but are not limited to local anesthetics, calcium channel blockers, beta blockers and cyclic antidepressants.
- Dystonic reactions may be associated with antipsychotic agents, neuroleptic agents, antiemetics, and other medications.
- Tricyclic toxicity can progress rapidly from alert mental status to cardiac arrest.
- 20% Fat emulsion should be drawn up into a syringe and administered IV/IO push.
- To prepare Hydroxocobalamin (Cyanokit), place the vial in an upright position; utilizing the supplied transfer spike, add 0.9% NaCl to the vial (200ml for 5gm vial) and fill to the marked line. Rock the vial for at least 60 sec. Infuse utilizing vented IV tubing.

Request a Recovery Coach now.
Don't wait!

TREAT
patient for overdose.

ASK
patient if Recovery Coach is wanted.

CALL NOW
and request a Recovery Coach
to meet patient at hospital.

415-8833

EMS provides lifesaving treatment, care, and transport.
Be part of the road back.

Anchor
Recovery Community Center
peer-to-peer support services

Recognition:
- Beta blockers/calcium channel blockers: bradycardia, hypotension, hyperglycemia [CCB].
- Tricyclic antidepressants/sodium channel blocking agents: altered mental status, seizures, hypotension, QRS widening, prolonged QTC, tall R in aVr.
- Opioids: altered mental status, pinpoint pupils, respiratory depression, and hypotension.
- Organosphosphates/nerve agents: salivation, lacrimation, urination, defecation, GI distress, emesis [SLUDGE].
- Anticholinergics: tachycardia, hyperthermia, dilated pupils, mental status changes.

E

- Routine patient care.
- Obtain history of ingestion/exposure including the substance ingested, route, quantity, time of ingestion, etiology (intentional, accidental, criminal), available medications/toxins in the home and the patient's past medical history and medications.
- For known or suspected **opioid overdose** with respiratory depression/apnea in patients <20 kg, administer NALOXONE 0.1 mg/kg IN (max dose 2 mg) every 3-5 min until adequate ventilation is restored. For patients ≥20 kg, administer NALOXONE 2 mg IN every 3-5 min until adequate ventilation is restored or 10 mg is administered.
- For known or suspected exposure to a **nerve or organophosphate agent,** manage per the Nerve Agent-Organophosphate Exposure Protocol.
- Consider consulting the **Regional Center for Poison Control and Prevention (800 - 222-1222)** for advice.
- For oral ingestions occurring less than one hour prior to EMS contact, contact **MEDICAL CONTROL** for authorization to administer ACTIVATED CHARCOAL 1 gm/kg (0.5 gm/lb) PO.
- Transport the patient to nearest appropriate Hospital Emergency Facility.

A
C

- For suspected **opioid overdose**, if NALOXONE has not been administered or if additive doses are required, NALOXONE 0.1 mg/kg IN/IV/IM (max dose 2mg) every 3-5 min until adequate ventilation is restored. For patients ≥ 20 kg, administer NALOXONE 2 mg IN/IV/IM every 3-5 min until adequate ventilation is restored or 10 mg is administered.
- For suspected **tricyclic antidepressant or other sodium channel blocking agent toxicity,** if seizures are present or the QRS is >0.09 sec, contact **MEDICAL CONTROL** for authorization to administer SODIUM BICARBONATE 1 mEq/kg IV (repeat every five min as indicated).
- For suspected **beta blocker or calcium channel blocker (CCB) toxicity**, manage as indicated per *Cardiac Dysrhythmia Protocol(s).* Contact **MEDICAL CONTROL** for authorization to administer GLUCAGON 0.1 mg/kg IV [max 5 mg] (may repeat X1 in 15 min) **and** CALCIUM CHLORIDE 20 mg/kg IV **or** CALCIUM GLUCONATE 60 mg/kg IV (may repeat X1).
- For suspected **cyanide toxicity**, contact **MEDICAL CONTROL** for authorization to administer (if available) HYDROXOCOBALAMIN 70 mg/kg IV (5 gm max) in 250 ml NS over 15 min (in cardiac arrest the dose may be given over a shorter period).
- For suspected **sympathomimetic/stimulant toxicity**, contact **MEDICAL CONTROL** for authorization to administer MIDAZOLAM 0.05 mg/kg IV/IN (may repeat X1 in 10 min) [do not administer if < 5 kg].

P

- For suspected **opioid overdose**, if NALOXONE has not been administered or if additive doses are required, consider NALOXONE 0.1 mg/kg IN/IV/IM (max dose 2mg), repeat dose at 3-5 min intervals until adequate ventilation is restored. For patients ≥20 kg, consider NALOXONE 2 mg IN/IV/IM, repeat at 3-5 min intervals until adequate ventilation is restored or 10 mg is administered.
- For suspected **tricyclic antidepressant or other sodium channel blocking agent toxicity:**
  - SODIUM BICARBONATE 50 mEq IV for seizures or the QRS is > 0.09 sec (may repeat every 5 min).
  - NOREPINEPHRINE 0.1-2 mcg/kg/min IV for hypotension.
  - 20% FAT EMULSION 1.5 ml/kg IV for profound hemodynamic compromise or cardiac arrest (may repeat X2).
- For suspected **beta blocker or calcium channel blocker (CCB) toxicity:**
  - *Cardiac Dysrhythmia Protocol(s)* as indicated
  - GLUCAGON 0.1 mg/kg IV [max 5 mg] (may repeat X1 in 15 min).
  - CALCIUM CHLORIDE 20 mg/kg **or** CALCIUM GLUCONATE 60 mg/kg IV (may repeat X1).
  - DOPAMINE 2-20 mcg/kg/min IV **or** EPINEPHRINE 0.1 mcg/kg/min IV for hypotension.
  - 20% FAT EMULSION 1.5 ml/kg IV for profound hemodynamic compromise or cardiac arrest (may repeat X2).
- For suspected **cyanide toxicity:** (if available) HYDROXOCOBALAMIN 70 mg/kg IV [5 gm max dose] in 250 ml NS over 15 min (in cardiac arrest the dose may be given over a shorter period).
- For **sympathomimetic/stimulant toxicity**, MIDAZOLAM 0.05 mg/kg IV/IN (may repeat X1 in 10 min) [do not administer if < 5 kg].
- For **dystonic reactions**, DIPHENHYDRAMINE 1 mg/kg IV IV [50 mg max dose] [do not administer if < 5 kg].
- For any **ingestion of highly lipid soluble medication or substance** with profound hemodynamic compromise or cardiac arrest, 20% FAT EMULSION 1.5 ml/kg IV (may repeat X2).

PEARLS:
- Consider polydrug overdose in all patients.
- Identification of toxidromes may help identify toxins and guide therapy.
- Consider cyanide toxicity in any patient with smoke inhalation with altered mental status and unexplained hypotension.
- Patients with suspected cyanide toxicity should be administered oxygen at the highest concentration available.
- Dystonic reactions may be associated with antipsychotic agents, neuroleptic agents, antiemetics, and other medications.
- Tricyclic toxicity can progress rapidly from alert mental status to cardiac arrest.
- 20% Fat emuslion should be drawn up into a syringe and administered IV/IO push.
- To prepare Hydroxocobalamin (Cyanokit), place the vial in an upright position; utilizing the supplied transfer spike, add 0.9% NaCl to the vial (200ml for 5gm vial) and fill to the marked line. Rock the vial for at least 60 sec. Infuse utilizing vented IV tubing.

Recognition:
- Patient with suspected or know exposure to a nerve or organophosphate agent with salivation, lacrimation, urination, defecation, GI distress, emesis [SLUDGE], muscle twitching, seizures, respiratory arrest), bradycardia, bronchorrhea, and/or bronchospasm.

E
- Utilize appropriate personal protective equipment (PPE).
- If the scene is determined to be unsafe, call for additional/appropriate resources and stage until the scene is safe.
- Ensure appropriate resources are available to perform decontamination.
- Consider activation of the CHEMPACK resources from the RI Department of Health through Incident Command and Regional Control.
- Obtain history of ingestion/exposure.
- Utilize the Multiple Patient Incident Protocol as indicated.
- Routine patient care.
- If DuoDote or Mark1 Antidote Kit(s) are available, treat as below:
- Consider consulting the **Regional Center for Poison Control and Prevention (800 -222-1222)** for advice.
- Transport the patient to nearest appropriate Hospital Emergency Facility.

DuoDote/Mark1 Antidote Kit Adult Dosing	
Symptom Severity	**Dosing**
Minor (respiratory distress + SLUDGE)	2 doses IM rapidly
Major (AMS, respiratory distress/ arrest)	3 doses IM rapidly

DuoDote/Mark1 Antidote Kit Pediatric Dosing (minor or major severity)	
Age ≤ 7 years	1 dose IM rapidly
Age 8-14 years	2 doses IM rapidly
Age ≥ 15 years	3 doses IM rapidly

A C
- Manage seizures with a benzodiazepine as per the age appropriate Seizure Protocol.

P
- Manage seizures with a benzodiazepine as per the age appropriate Seizure Protocol.
- For the patient with **minor symptoms** (respiratory distress + SLUDGE), if nerve antidote kits are unavailable or if additive doses are required, administer ATROPINE 2 mg (peds 0.02- 0.05 mg/kg) IV/IM every 5 min until symptoms resolve **and** (if available) PRALIDOXIME 600 mg (peds 15-25 mg/kg) IV over 30 min or IM.
- For the patient with **major symptoms** (minor + AMS or seizures), if nerve antidote kits are unavailable or if additive doses are required, administer ATROPINE 6 mg (peds 0.02- 0.05 mg/kg) IV/IM every 5 min until symptoms resolve **and** (if available) PRALIDOXIME 600 mg (peds 15-25 mg/kg) IV over 30 min or IM.

PEARLS:

- Follow local HAZMAT protocols for decontamination and use of personal protective equipment.
- Antidotal therapy should be started as soon as symptoms appear.
- Non-symptomatic patients should be decontaminated, moved to the cold zone and monitored for signs/symptoms of toxicity.
- If a MPI exhausts the supply of Nerve Agent Kits, use pediatric atropine formulation (if available). Use the 0.5 mg dose if patient is less than 40 pounds (18 kg), 1 mg dose if patient weighs between 40 to 90 pounds (18 to 40 kg), and 2 mg dose for patients greater than 90 pounds (>40 kg).
- Each nerve agent kit contains (roughly) 600 mg of pralidoxime (2-PAM) and 2 mg of atropine.
- For seizure activity, any benzodiazepine by any route is acceptable.
- For patients with major symptoms, there is no limit for atropine dosing.
- The main symptom that the atropine addresses is excessive secretions so atropine should be given until salivation improves.

CHEMPACK Container Resources

Item	Description	Quantity per Case	Cases
Mark-I autoinjector	Two-part autoinjector that contains Atropine [2mg] and Pralidoxime (2-PAM Chloride [600mg]	240	5
Atropine Sulfate 0.4mg/mL 20mL	Multi-dose vial of Atropine	100	1
Pralidoxime 1gm inj 20mL	Multi-dose vial of Pralidoxime	276	1
Atropen 0.5mg	Atropine autoinjector	144	1
Atropen 1.0mg	Atropine autoinjector	144	1
Diazepam 5mg/mL autoinjector	Diazepam autoinjector	150	1
Diazepam 5mg/mL vial, 10mL	Multi-dose vial of Diazepam	50	1
Sterile water 20cc	Vial of sterile water for injections	100	2

Recognition:
- Patient involved in fire or in a confined space/ poorly ventilated area with potential for the presence of carbon monoxide (combustion of carbon containing fuels or inadequate ventilation of natural gas).
- Headache, nausea, vomiting, dizziness, blurred vision, confusion, tachycardia, palpations, tachypnea, arrhythmia, hypotension, myocardial ischemia, seizures, change in mental status, unresponsiveness.

E

- Routine patient care.
- OXYGEN at the highest concentration possible.
- Treat as outlined in the table below:

SpCO/SpO2	Treatment
SpCO Unavailable	If the patient is asymptomatic, transportation may not be required if there are no other injuries or conditions requiring evaluation or treatment. Treat cardiac, respiratory or neurologic symptoms per appropriate protocol(s). Transport the patient to nearest appropriate Hospital Emergency Facility. If the patient is obtunded and CO poisoning is highly suspected and the patient does not meet criteria for transportation to a Level 1 Trauma Center or Burn Center, in consultation with MEDICAL CONTROL, consider transportation of the patient to a hyperbaric oxygen (HBO) capable facility (see Table 2 - Point of Entry in *Routine Patient Care Protocol*).
0-5%	No further monitoring of SpCO or treatment for CO exposure is required.
5-15% and SpO2 > 90%	If asymptomatic, no further treatment for CO exposure is required. Treat cardiac, respiratory or neurologic symptoms per appropriate protocol(s). Transportation is required.
> 15-25 or SpO2 <90	Treat cardiac, respiratory or neurologic symptoms per appropriate protocol(s). Transportation is required.
> 25	Treat cardiac, respiratory or neurologic symptoms per appropriate protocols. If the patient does not meet criteria for transportation to a Level 1 Trauma Center or Burn Center, in consultation with MEDICAL CONTROL, consider transportation of the patient to a hyperbaric oxygen (HBO) capable facility.

- For fire victims who are obtunded and or with unexplained hypotension, consider treating for cyanide toxicity per the age appropriate *Toxicological Emergencies Protocol.*
- If a source for CO is not identified, it should be recommended the patient have their home or work environment evaluated for the presence of CO.

A C P

- Cardiac monitoring is indicated in all symptomatic patients and acquire a multilead ECG and examin for evidence of myocardial ischemia.

PEARLS:

- Common symptoms of CO poisoning include headache, nausea, vomiting, altered mental status, neurologic impairments, visual disturbances, reddened eyes, chest pain, tachycardia, arrhythmia, seizures.
- Have a high index of suspicion for CO exposure in environments when there are multiple occupants/inhabitants with "flu like symptoms" and headache.
- Fetal hemoglobin has a greater affinity for CO than maternal hemoglobin. Patients who are pregnant or may be pregnant that have had exposure to CO should be advised SpCO levels measured in the field do not reflect fetal COHb levels and fetal COHb levels may be significantly higher and a hospital evaluation is recommended.
- Carbon monoxide and cyanide are both products of combustion and toxic levels of both often occur together. A low or absent detectable level of CO does not rule out the presence of cyanide.

Recognition:
- Blast explosion occurring in an open space (conventional blast) or enclosed/confined space (vehicle, building, bus, train).

E

- Ensure scene safety.
- If possible, determine the following:
  - Nature of incident (accidental v. intentional [terrorism]).
  - Nature of device (manufactured, improvised explosive device [IED], industrial equipment).
  - Nature of environment (conventional blast v. enclosed/confined space blast).
  - Potential for threat of particalization of hazardous materials.
  - Distance from blast, +/- protective barrier.
- Quantify and triage patients per the *Multiple Patient Incident Protocol(s)* as indicated.
- When operating at the scene of a known or suspected intentional incident, be cognizant for the threat of a secondary device.
- Routine patient care.
- Manage patient as indicated per age appropriate *Trauma and Burn Protocols*, *Crush Injury Protocol*, *Radiation Incident Protocol.*
- Transport the patient to the nearest appropriate Hospital Emergency Facility.

PEARLS:
- Use caution when moving patients involved in an intentional blast incident. Prior to moving any patient, observe for indications (presence of wires, lying on top of a package/backpack) that he/she is not connected to a secondary device.
- Primary blast injuries are produced by contact of the blast shockwave with the body. This results in stress and shear waves within tissues. Gas filled organs (colon, lungs, ears) are at particular risk. Common primary injuries include blast lung injury, bowel perforation, splenic/hepatic rupture, tympanic membrane rupture, ocular injuries, and concussion (concussion may be present without overt signs of head injury).
- Secondary blast injuries are produced by primary fragments (pieces of exploding weapon e.g. nails, ball bearings) and secondary fragments (environmental fragments e.g. glass). Common secondary injuries include penetrating injuries, traumatic amputations, and lacerations.
- Tertiary blast injuries occur when the blast wave propels individuals onto surfaces/ objects or objects onto individuals. Common secondary injuries include blunt injuries, crush syndrome, and compartment syndrome.
- Quaternary blast occur secondary heat and/or combustion. Common quaternary injuries include toxidromes from fuels/metals, burns, inhalation injury, and asphyxiation.
- Quinary blast injuries occur from specific additives (bacteria, radioactive material) "dirty bomb".
- High-order explosives (HE) produce a super-sonic over pressurization blast wave. HE examples include TNT, C-4, NTG, Symtex, and ammonium nitrate fuel oil (ANFO).
- Low-order explosive (LE) produce a sub-sonic explosion and therefor no over pressurization shock wave is created. LE examples include pipe bombs, gun powder and pure-petroleum based bombs (Molotov cocktails).
- The HE "blast wave" (over-pressure component) should be distinguished from "blast wind" (forced super-heated air flow). The latter may be encountered with both HE and LE.

PEARLS:

- "Blast lung injury" refers to the pulmonary sequelae of exposure of the lungs to an over pressurization blast wave. It is characterized by respiratory distress (dyspnea/apnea), bradycardia, and hypotension. Hemoptysis, hypoxia, and tension pneumothorax may also be present.
- Patients with blast lung injury may require early intubation, but overly aggressive positive pressure ventilation may worsen the injury. Use low tidal volumes and avoid hyperventilation in these patients. Volume overload may also worsen lung injury. IV fluids should be administered judiciously in the setting of lung injury (maximize preload without overload).
- The risk for blast lung injury is increased with closed/confined space blasts.

Recognition:

- Patient with radiation burn or exposure to radiation.

E

- Ensure scene safety.
- If the incident involves a blast, also manage per _Blast Incident and Injury Protocol_.
- If possible, determine the following:
  - Need for additional resources (hazardous material decontamination team etc.).
  - Exposure type: external irradiation, external contamination with radioactive material, internal contamination with radioactive material.
  - Quantification of exposure: generally measured in Grays/Gy (this information may be available from individuals on scene. Do not delay transportation to acquire this information).
- Quantify and triage patients as per the _Multiple Patient Incident Protocol_ as indicated.
- Routine patient care.
- Flush contact areas with NORMAL SALINE for 15 minutes.
- If there is ocular involvement, irrigate affected eye(s) with LACTATED RINGER'S **or** NORMAL SALINE for 15 min.
- If present, manage burn injuries as per age appropriate _Burn Protocol(s)._
- Identify and manage any secondary injuries as per appropriate _Trauma Protocols_.
- Transport the patient to the nearest appropriate Hospital Emergency Facility.

PEARLS:

- A patient who is contaminated with radioactive material (e.g. flecks of radioactive material embedded in their clothing and skin) generally poses minimal exposure risk to EMS providers.
- In general, trauma patients who have been exposed to or contaminated by radiation should be triaged and treated on the basis of the severity of their conventional traumatic injuries.

Section 5: Airway Protocols

Airway Management - Adult

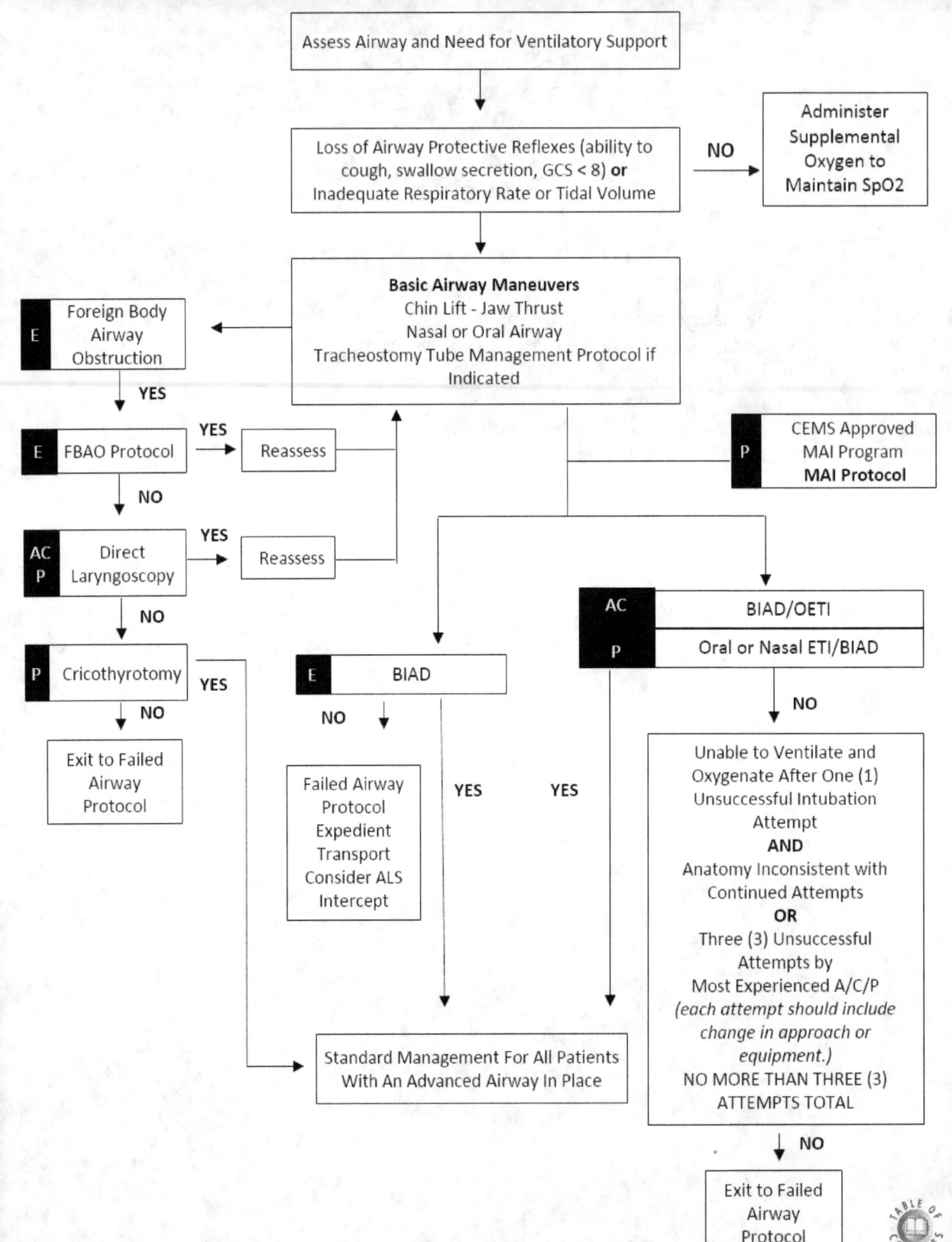

1. The risk of endotracheal intubation in the field should be weighed the option of transport.
2. Airway protective reflexes include the ability to cough and swallow secretions. Contrary to common EMS teaching, gag reflex is not an airway protective reflex.
3. The most important and most difficult to use airway device is the bag-valve-mask device. Providers must master the skill of BVM ventilation.
4. An intubation attempt is defined as the introduction of a laryngoscope blade, endotracheal tube or BIAD beyond the patient's lips or the insertion of an endotracheal tube into the patient's nasal passage.
5. A failed airway is defined as the inability to ventilate and oxygenate after one (1) unsuccessful intubation attempt **and** anatomy inconsistent with continued attempts **or** three (3) unsuccessful   attempts by the most experienced A/C/P **or** the inability to maintain oxygenation with BVM techniques and insufficient time to attempt alternative maneuvers.
6. The BURP (backwards, upward, rightward pressure over the larynx) maneuver may facilitate visualization of the glottic structures. Use of the Sellick maneuver (cricoid pressure) is not useful in aiding visualization of the glottic structures.
7. When possible, spontaneously breathing patients should be preoxygenated with high concentration oxygen via a nonrebreather mask combined with a nasal cannula (set at high flow) **or** with a BVM device attached to supplemental oxygen with a reservoir (preferably spontaneously breathing against mask seal without interposed positive pressure ventilation if rate and tidal volume allow) with a nasal cannula (set at high flow) **or** with CPAP. If used, the nasal cannula should be left in place during intubation.
8. Consider the following in airway assessment:

Difficult BVM Ventilation (MOANS):

M: Difficult mask seal due to facial hair, blood/secretions, anatomy, or trauma
O: Obesity or late pregnancy
A: Age (>55)
N: No teeth
S: Stiff or increased airway pressures (asthma, COPD, obesity, pregnancy)

Difficult Laryngoscopy (LEMON)

L: Look externally for anatomic distortions (small mandible, short neck, large tongue)
E: Evaluate using the 3-3-2 rule (mouth opening should accommodate 3 patient fingers, mandible to neck distance should accommodate 3 patient fingers, chin-neck junction to thyroid prominence should accommodate 2 patient fingers)
M: Mallampati (difficult to evaluate in the field or critically ill patient)
O: Obstruction (anatomic, pregnancy, obese)
N: Neck mobility (limited due to pathology, collar)

Difficult BIAD Insertion/ventilation (RODS):

R: Restricted mouth opening
O: Obstruction/obese/late pregnancy
D: Distorted or disrupted airway
S: Stiff lungs/increased airway pressure (asthma, COPD, obese, pregnant)

9. When intubating or placing a BIAW in a patient with a suspected cervical spine injury, maintain in-line cervical stabilization and remove the anterior section of the cervical collar to allow anterior mandibular mobility.
10. Nasotracheal intubation requires spontaneous ventilation. Colorimetric EtCO2 detection devices are not reliable for confirmation of a nasally placed endotracheal tube.
11. Paramedics should place a gastric tube in all intubated patients.
12. The use of continuous quantitative waveform in MANDATORY in all patient with an endotracheal tube in place.

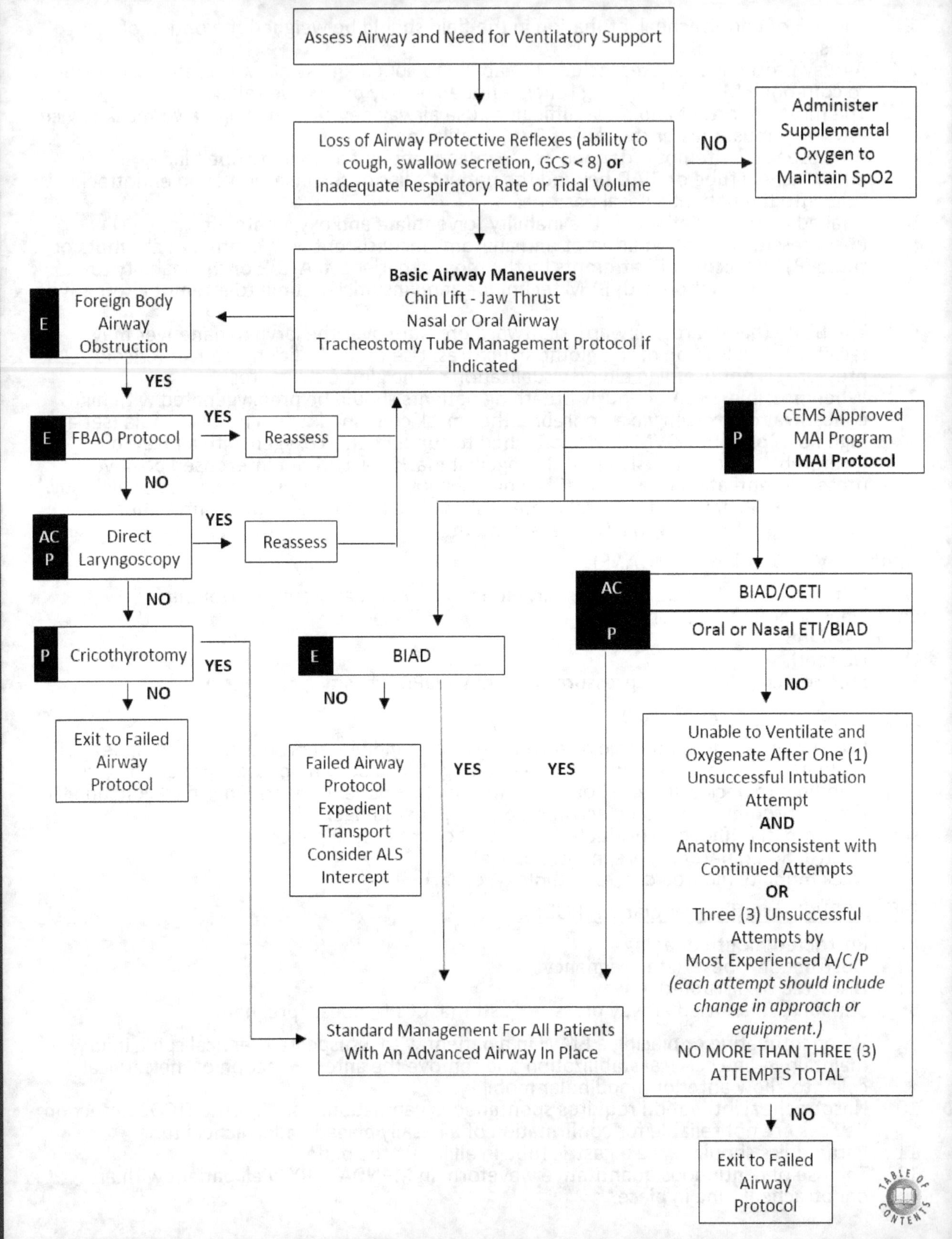

Assess Airway and Need for Ventilatory Support

Loss of Airway Protective Reflexes (ability to cough, swallow secretion, GCS < 8) **or** Inadequate Respiratory Rate or Tidal Volume

NO → Administer Supplemental Oxygen to Maintain SpO2

Basic Airway Maneuvers
Chin Lift - Jaw Thrust
Nasal or Oral Airway
Tracheostomy Tube Management Protocol if Indicated

E Foreign Body Airway Obstruction

YES

E FBAO Protocol

YES → Reassess

NO

AC P Direct Laryngoscopy

YES → Reassess

NO

P Cricothyrotomy

YES

NO

Exit to Failed Airway Protocol

P CEMS Approved MAI Program **MAI Protocol**

AC BIAD/OETI
P Oral or Nasal ETI/BIAD

NO

E BIAD

NO

YES

Failed Airway Protocol
Expedient Transport
Consider ALS Intercept

YES

Unable to Ventilate and Oxygenate After One (1) Unsuccessful Intubation Attempt
AND
Anatomy Inconsistent with Continued Attempts
OR
Three (3) Unsuccessful Attempts by Most Experienced A/C/P *(each attempt should include change in approach or equipment.)*
NO MORE THAN THREE (3) ATTEMPTS TOTAL

NO

Standard Management For All Patients With An Advanced Airway In Place

Exit to Failed Airway Protocol

1. The risk of endotracheal intubation in the field should be weighed the option of transport.
2. Airway protective reflexes include the ability to cough and swallow secretions. Contrary to common EMS teaching, gag reflex is not an airway protective reflex.
3. The most important and most difficult to use airway device is the bag-valve-mask device. Providers must master the skill of BVM ventilation.
4. An intubation attempt is defined as the introduction of a laryngoscope blade, endotracheal tube or BIAD beyond the patient's lips or the insertion of an endotracheal tube into the patient's nasal passage.
5. A failed airway is defined as the inability to ventilate and oxygenate after one (1) unsuccessful intubation attempt and anatomy inconsistent with continued attempts or three (3) unsuccessful attempts by the most experienced A/C/P or the inability to maintain oxygenation with BVM techniques and insufficient time to attempt alternative maneuvers.
6. The BURP (backwards, upward, rightward pressure over the larynx) maneuver may facilitate visualization of the glottic structures. Use of the Sellick maneuver (cricoid pressure) is not useful in aiding glottis visualization.
7. When possible, spontaneously breathing patients should be preoxygenated with high concentration oxygen via a nonrebreather mask combined with a nasal cannula (set at high flow) or with a BVM device attached to supplemental oxygen with a reservoir (preferably spontaneously breathing against mask seal without interposed positive pressure ventilation if rate and tidal volume allow) with a nasal cannula (set at high flow) or with CPAP. If used, the nasal cannula should be left in place during intubation.
8. In the pediatric patient, anticipate a higher and more anterior glottic opening.
9. Maintain the patient in the sniffing position. Do not hyperextend the neck. A towel may be required under the shoulders to elevate the torso relative the head in small infants.
10. A straight laryngoscope blade should be used in the patient < 3 years of age.
11. When intubating or placing a BIAW in a patient with a suspected cervical spine injury, maintain in-line cervical stabilization and remove the anterior section of the cervical collar to allow anterior mandibular mobility.
12. Paramedics should place a gastric tube in all intubated patients and pediatric patients receiving positive pressure ventilation via mask.
13. The use of continuous quantitative waveform in MANDATORY in all patient with an endotracheal tube in place.

Failed Airway - Adult

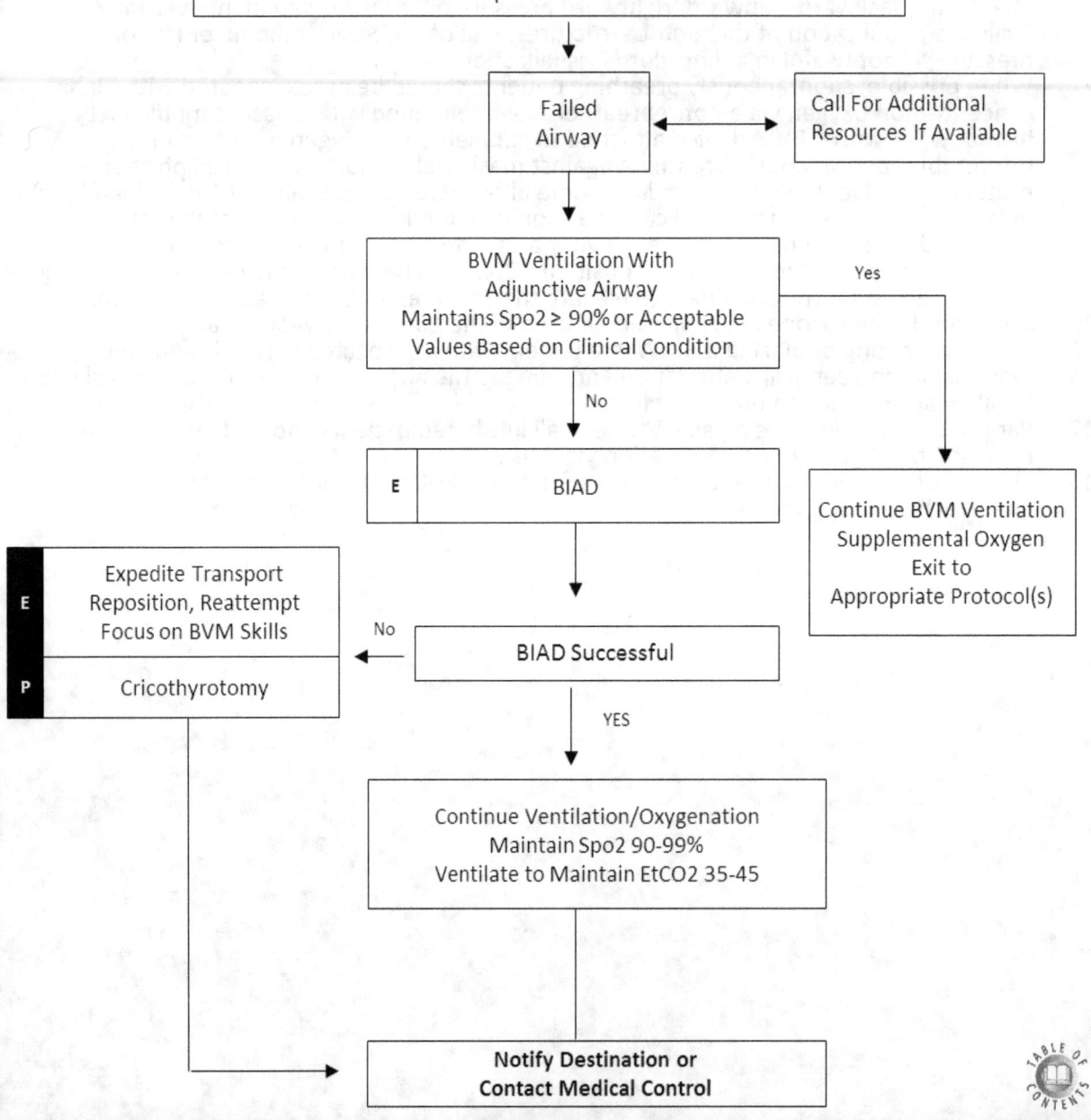

Unable To Ventilate and Oxygenate Adequately After One (1) or More
Unsuccessful Intubation Attempt
AND
Anatomy Inconsistent With Continued Attempts
OR
Three (3) Unsuccessful Attempts By
Most Experienced A/C/P
(each attempt should include change in approach or equipment)
NO MORE THAN THREE (3) ATTEMPTS TOTAL
OR
Inability to Maintain Oxygenation with BVM Techniques and Insufficient
Time to Attempt Alternative Maneuvers

Failed
Airway ⟷ Call For Additional
Resources If Available

BVM Ventilation With
Adjunctive Airway
Maintains Spo2 ≥ 90% or Acceptable
Values Based on Clinical Condition

Yes →

No

| E | BIAD |

Continue BVM Ventilation
Supplemental Oxygen
Exit to
Appropriate Protocol(s)

| E | Expedite Transport Reposition, Reattempt Focus on BVM Skills |
| P | Cricothyrotomy |

← No — BIAD Successful

YES

Continue Ventilation/Oxygenation
Maintain Spo2 90-99%
Ventilate to Maintain EtCO2 35-45

**Notify Destination or
Contact Medical Control**

1. A failed airway occurs when a provider begins a course of airway management by endotracheal intubation and identifies that intubation by that method will not succeed.
2. The most important way to avoid a failed airway situation is to identify patients with an expected difficult airway, difficult BVM ventilation, difficult BIAD insertion, difficult larygoscopy and/or difficult cricothyrotomy.
3. Improper positioning is commonly responsible for failed and difficult intubations in the field. Understanding that the prehospital setting I unique and possess many challenges, attempts to properly position the patient must be undertaken. The "sniffing position" or simply the head extended upon the neck are probably the best positions to facilitate airway management. The goal is to align the ear canal with the suprasternal notch in a straight line.
4. In obese or patients in late pregnancy, elevating the torso with blankets, pillows or towels will optimize patient position. This can also be accomplished by elevating the head of the ambulance cot.
5. Elevating or lowering the ambulance cot may be helpful. With the patient on the cot, raise the cot until the patient's nose is at the level of your umbilicus.
6. When intubating or placing a BIAW in a patient with a suspected cervical spine injury, maintain in-line cervical stabilization and remove the anterior section of the cervical collar to allow anterior mandibular mobility.
7. BVM ventilation can be maximized by the placement of 2 (two) nasopharyngeal airways and an oropharyngeal airway.
8. If the first intubation attempt is not successful, make an adjustment and consider:
   - Use of a different laryngoscope blade/video or other laryngoscope device if available.
   - Use of an endotracheal tube introducer
   - Downsizing the ETT
   - Apply the BURP maneuver.
   - Change head position or the level of the patient.

5.02 Pediatric Failed Airway - Pediatric

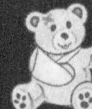

Unable To Ventilate and Oxygenate Adequately After One (1) or More Unsuccessful Intubation Attempt
AND
Anatomy Inconsistent With Continued Attempts
OR
Three (3) Unsuccessful Attempts By Most Experienced A/C/P
(each attempt should include change in approach or equipment)
NO MORE THAN THREE (3) ATTEMPTS TOTAL
OR
Inability to Maintain Oxygenation with BVM Techniques and Insufficient Time to Attempt Alternative Maneuvers

↓

Failed Airway ↔ Call For Additional Resources If Available

↓

BVM Ventilation With Adjunctive Airway
Maintains Spo2 ≥ 90% or Acceptable Values Based on Clinical Condition

- Yes → Continue BVM Ventilation
Supplemental Oxygen
Exit to Appropriate Protocol(s)
- No ↓

E | BIAD

↓

BIAD Successful

- No → E | Expedite Transport
Reposition, Reattempt
Focus on BVM Skills
P | Cricothyrotomy
→ Notify Destination or Contact Medical Control
- YES ↓

Continue Ventilation/Oxygenation
Maintain Spo2 90-99%
Ventilate to Maintain EtCO2 35-45

↓

Notify Destination or Contact Medical Control

TABLE OF CONTENTS

5.02 Pediatric Failed Airway - Pediatric

1. A failed airway occurs when a provider begins a course of airway management by endotracheal intubation and identifies that intubation by that method will not succeed.
2. The most important way to avoid a failed airway situation is to identify patients with an expected difficult airway, difficult BVM ventilation, difficult BIAD insertion, difficult larygoscopy and/or difficult cricothyrotomy.
3. Improper positioning is commonly responsible for failed and difficult intubations in the field. Understanding that the prehospital setting I unique and possess many challenges, attempts to properly position the patient must be undertaken. The "sniffing position" or simply the head extended upon the neck are probably the best positions to facilitate airway management. The goal is to align the ear canal with the suprasternal notch in a straight line.
4. In obese or patients in late pregnancy, elevating the torso with blankets, pillows or towels will optimize patient position. This can also be accomplished by elevating the head of the ambulance cot.
5. Elevating or lowering the ambulance cot may be helpful. With the patient on the cot, raise the cot until the patient's nose is at the level of your umbilicus.
6. When intubating or placing a BIAW in a patient with a suspected cervical spine injury, maintain in-line cervical stabilization and remove the anterior section of the cervical collar to allow anterior mandibular mobility.
7. BVM ventilation can be maximized by the placement of 2 (two) nasopharyngeal airways and an oropharyngeal airway.
8. If the first intubation attempt is not successful, make an adjustment and consider:
    *   Use of a different laryngoscope blade/video or other laryngoscope device if available.
    *   Use of an endotracheal tube introducer
    *   Downsizing the ETT
    *   Apply the BURP maneuver.
    *   Change head position or the level of the patient.

Recognition:

- Patient with tracheostomy tube experiencing respiratory distress, increased inspiratory time or increased resistance with use of manual resuscitation bag.

E

- Routine patient care.
- If there is no **tracheostomy tube in place:**
  - If a tracheostomy tube is available, allow a caregiver to insert the tracheostomy tube or place the tracheostomy tube into the stoma.
  - If there is continued respiratory distress following replacement of the tracheostomy tube and a caregiver trained tracheostomy tube management is present, have the caregiver suction the tracheostomy tube.
- If there **is a tracheostomy tube in place** and a caregiver trained in tracheostomy tube management is present, suggest the following:
  - If an obturator is in place, remove the obturator and reassess.
  - If a speaking valve or decannulating plug is in place, remove the valve or plug, suction the tracheostomy tube and reassess.
  - If an inner cannula (IC) is in place, remove and inspect the IC, clear/clean as needed and reassess after removal of the IC. If there is continued respiratory distress, the tracheostomy tube should be suctioned.
- If there is **continued respiratory distress** after the above, provide assisted ventilation as needed via the tracheostomy tube with a manual resuscitation bag and manage as indicated per the age appropriate *Respiratory Distress Protocol*.
- Transport the patient to the nearest appropriate Hospital Emergency Facility.

P

- If there is **no tracheostomy tube** in place:
  - Insert a tracheostomy tube (if available) or an appropriate size ETT into the stoma.
  - If there is continued respiratory distress following replacement of the tracheostomy tube/ETT, suction the tube.
  - Provide assisted ventilation via the tracheostomy tube with a manual resuscitation bag if indicated.
  - Manage as indicated per the age appropriate *Respiratory Distress Protocol.*
- If there is a **tracheostomy tube in place:**
  - If an obturator is in place, remove the obturator and reassess.
  - If a speaking valve or decannulating plug is in place, remove the valve or plug, suction the tracheostomy tube and reassess.
  - If an inner cannula (IC) is in place, remove and inspect the IC, clear/clean as needed and reassess after removal of the IC. If there is continued respiratory distress, suction the tracheostomy tube.
  - Provide assisted ventilation via the tracheostomy tube with a manual resuscitation bag if indicated.
  - Manage as indicated per the age appropriate *Respiratory Distress Protocol.*

PEARLS:

- Utilize family members/caregivers as a resource as they have knowledge and training regarding tracheostomy tube management.
- Utilize the patient's equipment if available and properly function.
- With certain tracheostomy tubes, patients cannot receive positive pressure ventilation without the inner cannula in place as the cannula contains the 15 mm adapter required to attach a bag valve device or ventilator circuit.
- In an emergency, patients with a dislodged tracheostomy tube that cannot be reinserted should be intubated.
- A tracheostomy is not considered mature until > 7 days.   Dislodgement of a tracheostomy tube from an immature stoma requires quick recognition as the stoma will begin to close.   Immediate treatment includes mask ventilation and endotracheal intubation.
- Always deflate the tracheal tube cuff prior to removal.
- To quickly estimate the appropriate size suction catheter to utilize with a given tracheostomy tube, multiply the tubes inner diameter (ID) X2 and utilize the next smallest size catheter (e.g. 6.0 mm ID X 2 = 12, next smallest catheter is 10 fr).
- Suctioning depth is usually 3-6 cm. Ask family member/caregiver. Do not force a suction catheter.   If unable to pass a suction catheter, the tracheostomy tube should be changed.
- Do not suction for > 10 sec at a time, pre-oxygenate before and in between attempts.
- The mnemonic **D**isplaced tube **O**bstructed tube **P**neumothorax **E**quipment failure is useful for trouble shooting problems with artificial airways.

Recognition:

- Patient ≥ 8 yo requiring emergent intubation.

1. Preparation:
  a. Perform an airway assessment to identify potential difficulties in management and to determine the appropriateness of administering a paralyzing agent.
  b. Identify anatomic landmarks related to performing cricothyrotomy; ensure the immediate availability of cricothyrotomy kit.
  c. Determine back-up options and plan (ensure availability of back up device(s) and resources).
  d. Ensure proper monitoring is in place (SpO2, EtCO2, ECG, NIBP).
  e. Rule out the presence of any contraindications to the use of succinylcholine.
  f. Assemble and check all required equipment as per standard routing for intubation.
  g. Ensure functional IV access (two preferred if possible).
  h. Review medication dosing and draw up medications (syringes should be labeled).
  i. Check availability/functionality of suction.
  j. Hemodynamic optimization (fluids or vasopressors) as clinically indicated.

2. Preoxygenation (as allowed by the patient's ability to cooperate and/or the clinical situation) may be performed by:
  a. Use of a standard nasal cannula set at ≥ 15 lpm (or as tolerated by the patient) and
  b. CPAP or BiPAP for 3 minutes or
  c. Non-rebreather mask set at ≥ 15 LPM for 3 minutes or
  d. A bag-valve-mask device with a one-way exhalation valve and an attached reservoir bag with attach to supplemental oxygen at ≥ 15 LPM with 8 positive pressure breaths (interpose positive pressure breaths only if the patient is unable to maintain an adequate rate and tidal volume) 3 minutes.
  e. The nasal cannula at a flow rate of ≥ 15 lpm should be left in place during direct laryngoscopy, intubation, and confirmation of endotracheal tube placement.

3. Paralysis with induction:
  a. Administer induction agent of choice (ETOMIDATE 0.2-0.3 mg/kg IV **or** KETAMINE 1.5 mg/kg)*
  b. Administer neuromuscular blocking agent of choice (SUCCINYLCHOLINE 1.5 mg/kg IV **or** ROCURONIUM 1.0 mg/kg [if succinylcholine is contraindicated])*

4. Position the patient to optimize visualization and minimize the rate of desaturation during laryngoscopy.

5. Place endotracheal tube with proof:
  a. Confirm mandibular flaccidity (45 seconds following the administration of succinylcholine, 60 seconds following the administration of Rocuronium) and perform direct laryngoscopy/intubation.
  b. Continually monitor the SpO2. If the SpO2 decreases to ≤92%, direct laryngoscopy/attempts at intubation should be aborted and the patient should be ventilated/oxygenated with a BVM device. The SpO2 nadir during this time should be documented.
  c. Confirm endotracheal tube placement by: 1) direct visualization of the ETT passing through the vocal cords; 2) presence of an EtCO2 waveform; and 3) presence of breath sounds.

P

6. **Post-intubation management:**
   a. Provide standard management for the patient with an advanced airway in place. Continuous quantitative capnography is mandatory.
   b. Provide analgesia, sedation and neuromuscular blockade as indicated per the age appropriate *Patient Comfort Protocol*.

PEARLS:

- **Execution of this protocol is restricted to Paramedics who have completed the perquisite CEMS and system Medical Director credentialing requirements.**
- Use of a sedating agent alone to facilitate intubation is considerably less desirable than utilizing a sedating agent in combination with a neuromuscular blocking agent. First attempt success rates are significantly increased utilizing a neuromuscular blocking agent.
- In the prearrest patient, a "crash sequence" utilizing only succinylcholine or Rocuronium may be indicated. These situations are rare and subsequent sedation must be provided when hemodynamically permissible.
- Medication assisted intubation includes rapid sequence intubation (the nearly simultaneous administration of a sedating agent and a short acting neuromuscular blocking agent to facilitate intubation) and delayed sequence intubation (procedural sedation with the procedure being oxygenation).
- In some instances, Ketamine may be used to facilitate preoxygenation (delayed sequence intubation) prior to RSI.
- Contraindications to succinylcholine include a patient or family history of malignant hyperthermia, burns, major crush injury injuries, denervation (spinal cord injury or CVA with paralysis) > 24 hrs (for the purposes of this protocol), myopathy, and preexisting hyperkalemia.
- The onset of action and duration of action of succinylcholine is 45 seconds and 6-10 minutes respectively. The onset of action and duration of action of Rocuronium is 60 seconds and 40-60 minutes respectively.
- The onset of action and duration of action of etomidate is 15-45 seconds and 3-12 minutes respectively. The onset of action and duration of action of ketamine is 45-60 seconds and 10-20 minutes respectively.
- Unless contraindicated, succinylcholine is the preferred agent for RSI.
- In both adults and pediatric patients, a repeat dose of succinylcholine may result in bradycardia requiring the administration of atropine sulfate.
- There is concern in the literature that Etomidate may inhibit endogenous cortisol production secondary to inhibition of 11-β-hydroxylase in the adrenal cortex. While there has been no large, randomized, prospective, adequately powered study that has demonstrated a significant negative effect associated with the administration of etomidate in the setting of sepsis, some providers avoid the use of etomidate in this setting.
- Very rarely the administration of succinylcholine may result in isolated masseter spasm causing mandibular rigidity (trismus) with limb muscle flaccidity. If this occurs and interferes with intubation, a non-depolarizing agents (Rocuronium 1 mg/kg) should be administered. The patient may require bag mask ventilation until complete muscle relaxation occurs.
- Fasciculations following the administration of succinylcholine are generally of no clinical significance.

Section 6: Special Situations

Multiple Patient Incident

Recognition:

- Incident resulting in ≥ 5 patients..

- If MPI is declared, upon arriving at the scene , immediate notification shall be made via the: Patient Tracking System's (PTS) MPI feature. All patients involved in the incident must be linked through PTS prior to arrival at designated hospital facility.
- Notify jurisdictional (local) communications center of MPI and request appropriate response (see Table 1 below) based on estimated number of patients/casualties.
- Use of the START/Jump Start Triage algorithm below is recommended (see PEARLS):

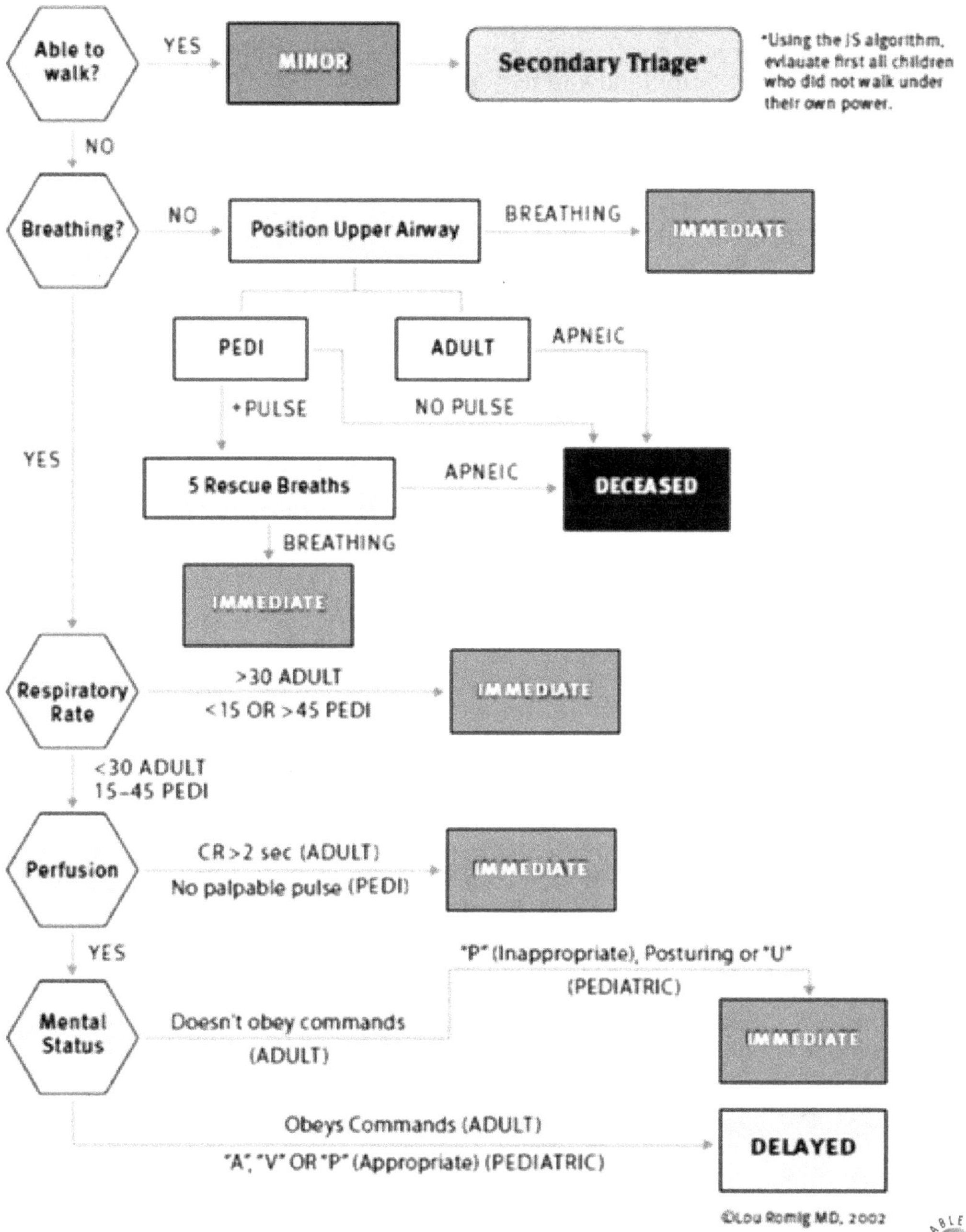
Able to walk?
YES
MINOR
Secondary Triage*
*Using the JS algorithm, evlauate first all children who did not walk under their own power.
NO
Breathing?
NO
Position Upper Airway
BREATHING
IMMEDIATE
PEDI
ADULT
APNEIC
+PULSE
NO PULSE
YES
5 Rescue Breaths
APNEIC
DECEASED
BREATHING
IMMEDIATE
Respiratory Rate
>30 ADULT
<15 OR >45 PEDI
IMMEDIATE
<30 ADULT
15-45 PEDI
Perfusion
CR>2 sec (ADULT)
No palpable pulse (PEDI)
IMMEDIATE
YES
Mental Status
"P" (Inappropriate), Posturing or "U"
(PEDIATRIC)
Doesn't obey commands
(ADULT)
IMMEDIATE
Obeys Commands (ADULT)
"A", "V" OR "P" (Appropriate) (PEDIATRIC)
DELAYED
©Lou Romig MD, 2002

E

TABLE OF CONTENTS

- If not done by the jurisdictional (local, system specific) communications center, EMS Incident Command should make primary declaration of a MPI to the HOST HOSPITAL via the RISCON 800 MHz on Wide Area 3.
- Utilizing the Jump Start Algorithm, first evaluate all children who did not walk under their own power.
- All infants who are patients in a MPI should be automatically triaged as "IMMEDIATE" or "red tag".
- Utilize PPE and request specialty resources as appropriate for the particular incident (hazardous materials, chemical, biological, nuclear, technical/confined space etc).
- Incident command should be established following jurisdictional (local) plans.
- Manage patients per specific protocols appropriate for injuries or illness.
- For **Nerve Agent** exposures, consider activation of CHEMPACK program through Incident Command and Regional Control

E

Table 1 - Southern New England Fire Emergency Assistance Plan - Tiered EMS Asset Levels

Tier	Primary EMS Assets	Additional Assets
Level 1	5 transporting	None
Level 2	5 transporting	None
Level 3	5 transporting	MCI Trailer
Level 4	5 transporting	None
Level 5	5 transporting	None
Level 6	5 transporting	MCI Trailer

PEARLS:

- The first arriving EMS unit should determine the nature of the incident, estimated number of patients and the need for specialty resources.
- Subsequently arriving EMS units should follow direction per the ICS and respond where and as directed by IC.
- EMS providers should maintain familiarity with the START/JUMP START or a similar algorithm and apply them if indicated with the understanding that they were developed to sort unresponsive patients with hemorrhagic shock from ambulatory patients in a first responder/BLS environment.
- In some incidents (fires, CO or chemical exposures etc.) ambulatory patients may have serous/life threatening injuries (e.g. inhalation injury).
- Municipal services should maintain a multiple patient (mass casualty) incident plan specific to their jurisdiction/region.
- Instances of lightning strike/electrocution the reverse triage rule applies. Patients without spontaneous circulation should be treated prior to those with.
-

Recognition:
- Patient with traumatic injuries or burns meeting the criteria for transportation to a Level I Trauma Center or a Verified Burn Center as outlined in the Multiple Trauma or Burn Protocols when the ground scene to door transportation time is ≥ 45 minutes.

- The use of HEMS may be considered in a multiple patient incident.
- Patients with an unmanageable airway or uncontrollable hemorrhage should be transported to the nearest appropriate Hospital Emergency Facility unless advanced life support or HEMS can intercept in a more timely fashion.
- Contact a HEMS provider (see Table I below) and determine availability and estimated time of arrival (ETA). **If the ETA of HEMS will extend the arrival of the patient to the intended receiving facility beyond that of ground transportation form the scene, the patient should be transported to the nearest appropriate Hospital Emergency Facility.**
- HEMS is not indicated for patients in cardiac arrest. If the patient experiences cardiac arrest after the request of HEMS, the patient should be transported to the nearest appropriate Hospital Emergency Facility. If the arrival of HEMS is imminent, the HEMS crew may be utilized for assistance in resuscitation and stabilization.
- Patients with exposure to hazardous materials must be decontaminated prior to the utilization of HEMS.
- Selection and preparation of a Landing Zone (LZ) should occur following local jurisdictional procedures. Radio communication with HEMS air assets should be via the designated LZ command.
-

Table 1 - Rhode Island Area HEMS Providers

Provider	Base Locations	Communication Center
UMass Memorial Life Flight	Worcester, MA	800-343-4354
Life Star	Hartford, Norwich, CT	800-221-2569
Boston Medflight	Bedford, Lawrence, Plymouth, MA	800-233-8998

Recognition:
- A patient with MOLST that are delineated on a R.I. Department of Health approved MOLST form (see reverse side of protocol) which has been signed by a MOLST qualified healthcare provider (MD/DO, NP, PA) and signed by the patient, the patient's decision maker, parent/guardian of a minor, or guardian as per R.I. General Law §23-4.11-3.1.
- A patient enrolled in the Comfort One Program with a Comfort One bracelet affixed to their person.

E

- Provide routine patient care and continue or initiate resuscitative efforts per age appropriate protocols until the patient's MOLST or Comfort One status have been verified.
- **For patients with MOLST**, verify the presence of an appropriately executed MOLST form:
  - Provide care as outlined on the patient's MOLST form.
  - If possible, determine if hospice or a home health agency is involved in the patient's care and provide appropriate notification.
  - Provide emotional support to the family.
- **For patients with Comfort One Status**, verify the presence of a Comfort One bracelet on one of the patient's extremities or in a sealed and closed bracelet on a neck chain. The bracelet must be intact and not defaced or damaged:
  - Do not initiate ventilator support, external chest compressions, advanced airway management (intubation, SGA), cardiac monitoring, establish IV access for resuscitative purposes, or provide electrical therapy (defibrillation, pacing).
  - Do provide airway suctioning, manual airway management and oxygen for comfort, positioning for comfort, analgesia for pain control, and hemorrhage control.
  - If possible, determine if hospice or a home health agency is involved in the patient's care and provide appropriate notification.
  - Provide emotional support to the family.

PEARLS:
- A patient with present capacity can, at any time, void the MOLST form or change their mind about their treatment preferences.
- A patient's healthcare decision maker may request to modify the MOLST based on the known desires of the patient or, if unknown, the patient's best interest.
- If the patient's medical records contain more than one MOLST form, the orders contained in the most recent MOLST form shall be followed unless the form is updated.
- Regardless of their mental or physical condition, a patient may revoke their Comfort One status by directly communicating their desire to EMS providers or another licensed health care provider or by physically destroying their Comfort One bracelet.
- Additionally, a patient's surrogate decision maker or another person in the patient's presence and at the direction of the patient may revoke a patient's Comfort One status by directly communicating this desire EMS providers or another licensed health care provider or by physically destroying the patient's Comfort One bracelet.
- In the event the patient or other qualified individual revokes a patient's MOLST or Comfort One status, the revocation must be documented in the EMS PCR and the health care provider(s) and staff directly responsible for the patient's care of the revocations.

Medical Orders for Life Sustaining Treatment (MOLST)

Follow these orders, then contact a MOLST-Qualified Health Care Provider. This is a **Medical Order Sheet** based upon the person's wishes in his/her current medical condition. Any section not completed implies full treatment. **This MOLST remains in effect unless revised.**

Patient's Last Name _____ Patient's First Name _____

Gender: ☐ M ☐ F Patient's Date of Birth __/__/__ Date/Time Form Prepared _____

A
CHECK ONE

CARDIOPULMONARY RESUSCITATION (CPR): *Person has no pulse and is not breathing.*
☐ **Attempt Resuscitation/CPR**   ☐ **Do Not Attempt Resuscitation/DNR** (Allow Natural Death)
- No defibrillator (including automated external defibrillators) should be used on a person who has chosen "Do Not Attempt Resuscitation."
- When not in cardiopulmonary arrest, follow orders in sections B and C.

B*
CHECK ONE

MEDICAL INTERVENTION: *Patient has a pulse and/or is breathing.*
☐ **Comfort Measures Only:** Use medication by any route, positioning, wound care and other measures to relieve pain and suffering. Use oxygen, suction and manual treatment of airway obstruction as needed for comfort. Use antibiotics only to promote comfort.
☐ **Limited Additional Interventions:** Includes care described above. Use medical treatment, antibiotics, and IV fluids as indicated. Do not intubate. May use non-invasive positive airway pressure. Generally avoid intensive care.
☐ **Full Treatment:** Includes care described above in Comfort Measures Only and Limited Additional Interventions, as well as additional treatment, such as intubation, advanced airway interventions, mechanical ventilation, and defibrillation/cardioversion as indicated.

C
CHECK ONE

TRANSFER TO HOSPITAL
☐ Do not transfer to hospital for medical interventions.   ☐ Transfer to hospital if comfort measures cannot be met in current location.

D
CHECK ONE

ARTIFICIAL NUTRITION (For example a feeding tube): *Offer food by mouth if feasible and desired.*
☐ No artificial nutrition   ☐ Defined trial period of artificial nutrition
☐ Long-term artificial nutrition, if needed   ☐ Artificial nutrition until not beneficial or burden to patient

E
CHECK ONE

ARTIFICIAL HYDRATION: *Offer fluid/nutrients by mouth if feasible and desired.*
☐ No artificial hydration   ☐ Defined trial period of artificial hydration
☐ Long-term artificial hydration, if needed   ☐ Artificial hydration until not beneficial or burden to patient

F

ADVANCE DIRECTIVE (if any): *Check all advance directives known to be completed.*
☐ Durable Power of Health Care   ☐ Health Care Proxy   ☐ Living Will   ☐ Documentation of Oral Advance Directive

Discussed with:
☐ Patient   ☐ Health Care Decision Maker   ☐ Parent/Guardian of Minor   ☐ Court-Appointed Guardian   ☐ Other: _____

G

SIGNATURE OF MOLST-QUALIFIED HEALTHCARE PROVIDER (Physician, RNP, APRN, or PA)
My signature below indicates to the best of my knowledge that these orders are consistent with the person's medical condition and preferences.

Signature (required) _____ Phone Number _____ Date/Time __/__/__

Print Name _____ Rhode Island License # _____

SIGNATURE OF PATIENT, DECISION MAKER, PARENT/GUARDIAN OF MINOR, OR GUARDIAN
By signing this form, the patient or legally-recognized decision maker acknowledges that this request regarding resuscitative measures is consistent with the known desires of, and with the best interest of, the individual who is the subject of the form.

Signature (Required) _____ Phone Number _____ Relationship (if patient, write self) _____

Print Name and Address _____

SEND MOLST FORM WITH PERSON WHENEVER TRANSFERRED OR DISCHARGED.

**HIPAA PERMITS DISCLOSURE OF MOLST TO OTHER HEALTH CARE PROFESSIONALS AS NECESSARY.
MOLST IS VOLUNTARY. NO PATIENT IS REQUIRED TO COMPLETE A MOLST FORM.**

Review and Renewal of MOLST Orders on This MOLST Form (this MOLST form remains in effect unless another MOLST form is executed.)

The MOLST-Qualified Health Care Provider may review the form from time to time as the law requires, and also:

- If the patient moves from one location to another to receive care; or
- If the patient has a major change in health status (positive or negative); or
- If the patient or other decision-maker changes his/her mind about treatment.

Date/Time	Reviewer's Name and Signature	Location of Review (e.g., Hospital, Nursing Home, Provider's Office, Patient's Residence)	Outcome of Review
			☐ No change ☐ Form voided, new form completed ☐ Form voided, no new form
			☐ No change ☐ Form voided, new form completed ☐ Form voided, no new form
			☐ No change ☐ Form voided, new form completed ☐ Form voided, no new form

Directions for MOLST-Qualified Health Care Providers Completing MOLST

- Must be completed by a MOLST-Qualified Health Care Provider based on patient preferences and medical indications. A MOLST-Qualified Health Care Provider is defined as a physician, nurse practitioner, advanced practice registered nurse, or a physician assistant.

- MOLST must be signed by a MOLST-Qualified Healthcare Provider (physician, nurse practitioner, advanced practice registered nurse, or physician assistant) and the patient/decision maker to be valid. Verbal orders are acceptable with follow-up signature by provider in accordance with facility/community policy and documentation that there was discussion with the patient or the patient's advocate about discontinuing the MOLST order.)

- This is the ONLY MOLST FORM that is acceptable for completion in Rhode Island. Do not make your own MOLST form. Photocopies and faxes of signed MOLST forms are legal and valid.

- Any incomplete section of the MOLST form implies full treatment for that section.

*Section B:

- When comfort cannot be achieved in the current setting, the person, including someone with "Comfort Measures Only," should be transferred to a setting able to provide comfort (e.g., treatment of a hip fracture)

- IV medication to enhance comfort may be appropriate for a person who has chosen "Comfort Measures Only".

- Non-invasive positive airway pressure includes continuous positive airway pressure (CPAP), bi-level positive airway pressure (BiPAP), and bag valve mask (BVM) assisted respirations.

- Treatment of dehydration prolongs life. A person who desires IV fluids should indicate "Limited Interventions" or "Full Treatment."

Modifying and Voiding MOLST

- A patient with capacity can, at any time, void the MOLST form or change his/her mind about his/her treatment preferences by executing a verbal or written advance directive or a new MOLST form.

- To void MOLST draw a line through Sections A through E and write "VOID" in large letters. Sign and date the line.

- A health care decision maker may request to modify the orders based on the known desires of the individual or, if unknown, the individual's best interests.

DEFINITIONS

"Medical orders for life sustaining treatment" or "MOLST" means a voluntary request that directs a health care provider regarding resuscitative and life-sustaining measures. Rhode Island General Laws §23-4.11-2 (10).

"Qualified patient" means a patient who has executed a declaration in accordance with this chapter and who has been determined by the attending physician to be in a terminal condition. Rhode Island General Laws §23-4.11-2 (16).

"Terminal condition" means an incurable or irreversible condition that, without the administration of life sustaining procedures, will, in the opinion of the attending physician, result in death." Rhode Island General Laws §23-4.11-3.1 (20).

This form is approved by the Rhode Island Department of Health. For more information or a copy of the form, visit **www.health.gov**

**SEND MOLST FORM WITH PERSON WHENEVER TRANSFERRED OR DISCHARGED.
Rhode Island General Laws §23-4.11-3.1 authorizes this MOLST form.** (Rev. 9-2013)

Recognition:
- A patient ≥ 16 yo with a present mental capacity who wishes to refuse patient care or transportation by EMS providers for themselves **or**
- A minor parent with a present mental capacity who wishes to refuse patient care or transportation by EMS providers for their child **or**
- A patient < 16 years of age whose parents or legal guardian with a present mental capacity wishes to refuse patient care or transportation by EMS providers for their child or a child in their guardianship **or**
- A legally emancipated minor.

- A refusal of patient care is considered valid if the following three components are established:
  - **Competence:** Patients meeting the criteria in the Recognition section above are considered by R.I. General Law to be competent to consent to or refuse care. Parents and legal guardians, as described above, who are on scene may refuse consent to or refuse care on behalf of a minor child.
  - **Capacity:** The patient or legal guardian must demonstrate present mental capacity as assessed below.
  - **Informed refusal:** A patient must be fully informed about his or her medical/traumatic condition and the risks refusal and the benefits of treatment/transport in accordance with their presenting complaint.
- If the patient is exhibiting suicidal or homicidal ideations, contact the appropriate jurisdictional law enforcement agency.
- Perform an assessment of the patient's medical/traumatic condition, and, to the extent permitted by the patient, a physical examination including vital signs. This assessment, or the patient's refusal of assessment must be fully documented on the PCR.
- Attempt to identify any patient/guardian perceived barriers to treatment/transport and make reasonable efforts to address these barriers. This may include, but is not limited to the offer of transportation to a licensed healthcare facility not recommended by protocol. These offers should be made only for the purpose of facilitating additional evaluation and/or treatment which would otherwise be refused.
- Consider consulting **MEDICAL CONTROL** for assistance in facilitating patient acceptance of treatment or transportation. It may be helpful to have the patient or the patient's guardian speak directly with the **MEDICAL CONTROL** physician. If **MEDICAL CONTROL** is consulted, it should be included in the PCR (including the time of the consult and the name of the physician).
- Inform the patient/guardian of the risks of refusal and the benefits of treatment/transport in accordance with their presenting complaint. It should be explained that the list of risks described is not comprehensive due to the diagnostic limitations of the pre-hospital environment and that their refusal may result in worsening of their condition, serious disability or death.
- Assess the patient for their ability to demonstrate present mental capacity by determining the following. If the answer to all three is yes, present mental capacity is affirmed:
  - Does the patient/guardian understand the illness or injury and the benefits of evaluation, treatment and/or transportation by EMS? and
  - Does the patient/guardian understand the consequences (including death) of not seeking evaluation, treatment and/or transport by EMS? And
  - Does the patient understand alternatives to immediate evaluation, treatment and/or transportation by EMS?

E

- A competent patient who is determined to have present mental capacity that meets the following criteria may refuse evaluation, care or transport by EMS:
  - The refusal is solely initiated by the patient, not suggested/prompted by EMS providers;
  - The patient is oriented to person, place, time and situation;
  - There is no evidence of altered consciousness resulting from head trauma, medical illness, intoxication, dementia, psychiatric illness or other etiologies;
  - There is no evidence of impaired judgement from alcohol or drug influence;
  - There is no language communication barriers; and
  - There is no evidence or admission of suicidal ideation resulting in any gesture or attempt at self-harm and no verbal or written expression of suicidal ideation regardless of any apparent inability to complete a suicide attempt.
- Document the refusal of care/transportation by having the patient sign (or, in the case of a minor patient, the minor patient's parent, legal guardian, or authorized representative) sign a refusal of care statement on the PCR or a standalone, service specific refusal of care form. Documentation should also include, when possible, a signature by a witness, preferably a competent relative, friend, police officer, or impartial third party.
- Advise the patient/guardian that they should seek immediate medical care at an Emergency Department and that they may call 911 at any time if their condition changes or worsens or if they wish to be transported.
- Provide documentation on the PCR supporting the presence of mental capacity and specific information provided to the patient/guardian regarding their condition and risks associated with the refusal of evaluation, treatment and/or transportation by EMS.
- If a patient refuses to sign a refusal of care statement, provide documentation on the PCR regarding the situation under which the patient refused to sign.

E

Recognition:
- Patients undergoing interfaculty transport should be classified and aligned with transport resources appropriate for their needs.

  - The following classification should be utilized:

  Class A: A patient who is clearly and completely stable with a minimal potential to decompensate during transport. Example: a patient with no IV who being transported for diagnostic testing.

  Class B: A stable patient as above with IV fluids infusing without additive medications. Example: a patient with maintenance IV fluids running.

  Class C: A patient who has been stabilized as much as possible, but may compensate during transport. Patient has no medications or technology beyond the scope of practice of the highest licensed EMS provider in attendance. Example: a cardiac patient with heparin and IV nitroglycerin infusing.

  Class D: A patient with an acute injury or illness who may become unstable during transport and requires medications or technology not within the scope of practice of the highest licensed EMS provider in attendance and/or may develop complications requiring interventions beyond the scope of practice of the highest licensed EMS provider in attendance. Example: a patient receiving 2 or more vasopressors and who is receiving aortic counterpulsation therapy with and intra-aortic balloon pump.

- The transferring medical provider is responsible for ensuring the patient is aligned with appropriate transport personnel and technology resources based on their classification.
- The following details appropriate transport resources:

Table 1		
Class	Staffing	Vehicle
A	EMT	BLS
B	A/C, P	ALS
C	P	ALS
D	A/C/P + MD NP. PA, RN	ALS

- In the event where an ALS unit is required and the sending facility has made a reasonable effort to utilize an ALS unit and is unable to access one due to time constraints or the patient's condition, a BLS unit may be utilized, providing that appropriate supplies, equipment (refer to Table 2 in reverse), and verbal/written orders have been provided and staff qualified to provide expected care are to accompany the patient.
- For class A, B, or C transfers the highest licensed EMS provider will assume ultimate authority for patient treatment within the scope of the appropriate RI Statewide EMS Protocols.
- For class D transfers, the ultimate authority rests with the sending physician or licensed independent provider. Unless a physician is present during transport, authority shall be shared and care will be provided in a collaborative nature. The accompanying staff shall be responsible for care beyond that of the scope of the EMS providers present. The EMS providers present retain their authority for EMS care under their scope as defend in RI EMS Regulations and protocols. If questions arise or guidance is needed, contact the sending provider.

E

- All patients intubated with an endotracheal tube must have continuous waveform capnography monitoring in place during transfer/transport.

Table 2 - Equipment for Transfer Using a BLS Vehicle	
1 - Bag Normal Saline (1000 ml) 1 - Bag Lactated Ringer's (1000ml) 1- 60 drop IV administration set 1 - 10 or 20 drop IV administration set IV catheters (3 ea 14g, 16g, 18g, 20g) Manual defibrillator w/monitor Adenosine Amiodarone Calcium chloride or gluconate Dextrose 10% (250 ml bag) Dextrose 25% and 50% Midazolam Diphenhydramine (PO) Diphenhydramine (injectable) Dopamine Epinephrine 1:1:000 Epinephrine 1:10,000 Fentanyl Furosemide	Glucagon Hydrocortisone (injectable) Levophed Lidocaine Magnesium sulfate Sodium bicarbonate Naloxone Nitroglycerine (tabs or lingual spray)

P

- In addition the pharmacologic agents contained within the RI EMS Statewide EMS Protocols, paramedics may transport any patient receiving heparin and stroke patients receiving tissue plasminogen activator, nicardipine and labetalol.
- Paramedics may transport patients receiving IV antibiotics, any variation of crystalloid IV solutions and IV infusions of insulin, octreotide, vasopressin, or those containing electrolytes (potassium, magnesium, calcium, potassium phosphate, sodium phosphate). Potassium infusions >10 meq/hr (max 20 meq/hr) require continuous ECG monitoring.
- All infusions, unless otherwise exempted in these protocols must be delivered by electronic infusion pump.
- Paramedics who have completed a CEMS and service medical director approved blood and blood component competency program may transport patients receiving blood and blood products.
- Paramedics who have completed a CEMS and service medical director approved chest drainage system competency program may transport patients requiring a chest drainage system.

Recognition:
- Patient in the custody of law enforcement or correctional personnel.

E

- Routine patient care.
- For patients with a compliant/evidence of traumatic injury or medical illness, manage per appropriate protocol(s).
- For patients with a presentation consistent with excited delirium, manage per the *Excited Delirium Protocol.*
- If the patient has been exposed to **aerosolized oleoresin capsicum (OC, Cap-Stun®):**
  - Remove patient to fresh air if possible/appropriate.
  - Remove contaminated clothing.
  - Discourage patient from rubbing their eyes or other affected areas.
  - Reassure the patient the effects are temporary and should abate within 45 min.
  - Have the patient blow their nose to aid in dislodging residual irritant.
  - Irrigate the face/eyes with NORMAL SALINE or water.
  - If the patient is experiencing dyspnea or wheezing, manage per the age appropriate *Respiratory Distress Protocol.*
  - Patients with without dyspnea or wheezing with a history of asthma or COPD should be monitored for a period of 20 minutes.
- For patients that have been subjected to a **conductive energy weapon (CEW) [Taser©]:**
  - If CEW probes are embedded in sensitive/vulnerable areas (above the level of the clavicles, genitalia, female breasts) or if there is suspicion of probe penetration into bone or vasculature, the probes should not be removed by EMS providers. In this case, the patient should be transported for evaluation and probe removal.
  - CEW probes embedded subcutaneously in non-sensitive areas of skin may be removed by EMS providers as follows:
    1. Ensure wires are disconnected from the CEW.
    2. Stabilize the skin around the embedded probe using the non-dominant hand.
    3. Grasp the probe by the metal body using the dominant hand.
    4. Remove the probe in a single quick motion.
    5. Wipe the wound with chlorohexadine or alcohol and apply a dressing.
    6. Treat probes as a sharps hazard and dispose of properly (law enforcement may be required to take possession of the probes).
    7. Patients should be referred accordingly for follow up based on their tetanus immunization status.
    8. Patients who have been exposed to > 3 CEW cycles, to the effects of >1 CEW or a continuous CEW cycle of > 15 seconds should have ECG monitoring and be transported for evaluation.
- Transport patient to the nearest appropriate Hospital Emergency Facility.

PEARLS:
- Patients in police custody retain the right to participate in in decision making regarding their medical care and may request medical care of EMS.
- Patients restrained by law enforcement devices must be transported accompanied by law enforcement/correctional personnel in the patient compartment who is capable of removing the device(s). EMS providers utilize restraining devices in accordance with the *Patient Restraint Procedure Protocol*, law enforcement personnel may follow behind the ambulance during transport, if there are no safety concerns and the arrangement is agreeable to both EMS and law enforcement personnel on scene.
- If restraints are applied by law enforcement personnel, EMS providers are required to follow the *Patient Restraint Procedure Protocol* with regard to monitoring and documentation.
- Never position or transport any restrained patient in such a way that could negatively impact the patient's respiratory or circulatory status.
- If an asthmatic patient is exposed to OC and released to law enforcement, all parties should be advised to immediately contact EMS if wheezing or respiratory distress develop.

Recognition:
- Patient requiring specialized healthcare needs.

- Routine patient care.
- Attempt to contact the person most knowledgeable about the patient's specialized health care needs in order to obtain advice during the care and transport process.
- Determine if an Emergency Care Plan (ECP) is present. The ECP must include:
  - Patient identification, including photograph;
  - A brief description of the patient's specialized care needs;
  - Instructions for care in anticipated emergency situations;
  - Reference numbers for further information; and the
  - Filing and effective date from the Department of Health.
- If an ECP IS present, follow the ECP and transport the patient as indicated.
- If no ECP present, and the patient is attached to portable special medical equipment that appears to be working properly, transport it with the patient.
- If the equipment is either too large to transport or does not appear to be working properly, consult **MEDICAL CONTROL** for guidance. It may be necessary to as safely as possible remove the equipment from the patient and provide alternative support as indicated.
- If the patient has a specialized health care need not related to equipment, follow the instructions of the ECP.
- If there is an ECP, it should be kept with the patient during transport.
- If available, the person most knowledgeable about the patient's specialized health care needs should accompany the patient during transport.
- Transport the patient to nearest appropriate Hospital Emergency Facility.

E

Recognition:

- Pulseless and apneic patient undergoing resuscitative efforts meeting the criteria below.

    1. Paramedic level providers may terminate resuscitative efforts in the field if ALL of the following criteria are met:

        a. Patient is ≥ 16 years of age (or a patient less ≤ 16 if the parent(s) is/are agreeable);
        b. Persistent asystole or agonal rhythm (rate ≤ 20) is present and no reversible causes have been identified and high quality, minimally interrupted CPR has been performed for a <u>minimum</u> of 30 minutes **or** for a <u>minimum</u> of ≥ 40 minutes when other underlying ECG rhythms are present;
        c. NO SHOCKABLE RHYTHM WITHIN LAST 20 MINUTES
        d. The airway has been successfully managed with verification of device placement (including continuous waveform capnography). Acceptable management techniques include endotracheal intubation (oral or nasal), BIAD, or cricothyrotomy;
        e. IV or IO access has been achieved;
        f. Rhythm appropriate pharmacologic and electrical therapy has been administered according to applicable protocols;
        g. Patient is not profoundly hypothermic;
        h. All possible reversible etiologies have been considered and managed.

P

    2 When considering TOR, consider allowing the family to be present during resuscitative efforts as appropriate.
    3 MEDICAL CONTROL may be consulted at any time when considering TOR. If MEDICAL CONTROL was consulted, note and document the name of the physician consulted.
    4 Follow protocol *06.09 Deceased Persons*.

PEARLS:

- Resuscitative efforts shall not be terminated in pt. with a shockable rhythm (VF/VT) unless authorized by MEDICAL CONTROL

Recognition:

- Patient encountered by EMS providers who meets criteria for biological death.
- Patient with MOLST/Comfort One in place who is pulseless and apneic.
- Patient with other approved advanced directive requiring no CPR be administered who is pulseless and apneic.
- Patient for whom resuscitative efforts are terminated on scene by Paramedic level EMS providers.

1. Once a determination of death has been made by EMS providers, notify the law enforcement agency having jurisdiction.

2. If the death is of a suspicious nature:
   a. Follow the direction of law enforcement personnel;
   b. Limit the number of EMS providers at the immediate scene to preserve the integrity of the scene; and
   c. Provide any requested or relevant information to law enforcement personnel.

3. If the determination of death has been made following the termination of resuscitation (TOR) in the field, leave all lines, tubes, and other therapeutic devices (electrical therapy pads, ECG electrodes etc.) in place until it is known whether the case will be accepted by the Office of the State Medical Examiner (OSME). If any devices were removed during resuscitative efforts for any reason, tape them across the chest of the patient. Do not place sharps under tape but rather note the devices in writing on the tape.

4. Death occurring from any of the following must be referred to the OSME and the law enforcement agency having jurisdiction is responsible for notifying the OSME:
   a. Death is due to, or there is a suspicion of accident, homicide, suicide, or trauma of any nature;
   b. Death is sudden in a public place;
   c. Death is from a drug or toxic substance;
   d. Death is sudden and unattended (patient without a primary care physician who apparently dies of natural causes aka "natural death);
   e. Death is related to a job, work place or environment;
   f. Death occurs during or immediately after surgery or diagnostic or therapeutic procedure.

5. If the death is an attended death (the patient has an active primary care physician and has died of apparently natural causes) and the death does not meet any of the criteria in #4 above:
   a. A reasonable effort should be made by EMS to notify the patient's primary care physician of the death. It is preferred (but not mandatory) to communicate directly with the primary care physician prior to releasing the body.
   b. Document any communication with the patient's primary care physician in the EMS patient care report (PCR) and provide the name of the physician who will be signing the death certificate to law enforcement.
   c. If unable to make contact with the patient's primary care physician, provide the name of the physician and his/her contact information to law enforcement personnel.
   d. The body may be released to a funeral home.

E

6. In certain circumstances (public area, highway or roadway), law enforcement may authorize the movement of a deceased person from the immediate scene.   Once a determination of death has been made at the scene, deceased persons are not to be transported to any facility (hospital, funeral home, OSME) in an ambulance.

7. EMS providers should:
   a. Maintain respect for the deceased and their family;
   b. Provide emotional support for bystanders and family members as appropriate; and
   c. Assist family members in contacting other family members or friends for emotional support.

E

Section 7:   Procedure Protocols

Indication:
- Acute decompensated heart failure/cardiogenic pulmonary edema.
- Respiratory distress/hypoxic respiratory failure associated with asthma/COPD, pneumonia, or near drowning.

Contraindications:
- Respiratory arrest/agonal breathing
- Patient unable to maintain airway
- GCS < 8 or patient unable to follow commands
- Vomiting or active upper GI hemorrhage
- Facial fractures or deformities prohibiting adequate mask seal
- Pneumothorax

Background:
Continuous positive airway pressure (CPAP) is a method of delivering oxygen via a positive pressure. CPAP raises inspiratory pressure above atmospheric pressure and maintains the pressure during exhalation. CPAP is easily delivered in the field. CPAP has been shown to decrease to the need for endotracheal intubation in cardiogenic pulmonary and is considered a first line treatment for cardiogenic pulmonary edema. CPAP is also useful in the management of respiratory distress and hypoxic respiratory failure associated with asthma, COPD, pneumonia and near drowning.

E

Procedure:
1. Ensure all necessary equipment is available, assembled and functional. Select appropriate mask size for patient.
2. Connect CPAP device to oxygen source and be sure oxygen is flowing.
3. Explain procedure to the patient.
4. Consider placing a nasopharyngeal airway.
5. With oxygen flowing, place the mask over the patient's mouth and nose and secure the mask with the provided straps starting with the lower strap until there is minimal or no air leak.
6. Select appropriate liter flow or adjust setting on the CPAP device to delivery desired level of PEEP. Start at zero and titrate upward based on pathology and patient response to a max of 10 cmH20 (5-10 cmH20 for pulmonary edema, near drowning, aspiration, pneumonia, 3-5 cmH20 for asthma or COPD).
7. Evaluate patient response by assessing breath sounds, SpO2, vital signs and general appearance.
8. If the administration of nebulized medication (albuterol) is required, follow device manufactures instructions.
9. If the patient's condition stabilizes, maintain CPAP for transportation. Provide the receiving facility with early notification of a patient requiring transition to hospital CPAP.
10. Discontinue CPAP and provide airway management/ventilator support as indicated if any of the following occur:
    - A decreased in the SpO2 (worsening hypoxia) from the initial reading when CPAP was applied;
    - Agonal breathing/respiratory arrest;
    - Decreased LOC (GCS < 8); or
    - Pneumothorax.
11. Document CPAP settings, patient's response and serial SpO2/vital signs/ capnographic readings.
12. Paramedics may administer MIDAZOLAM 2.5 mg IV/IN (may repeat X1 in 5 min) **or** LORAZEPAM 0.5-1 mg IV (may repeat X1 in 10 min) to facilitate the delivery of CPAP or if anxiolysis is required and the SBP is >100.

PEARLS:
- Close patient monitoring is required during the application of CPAP.
- Patients receiving CPAP may require coaching. Patients should be encourage to breathe slowly and deeply. The patient should be encourage to allow forced ventilation to occur.

Indication:
- Foreign body airway obstruction as evidenced by display of the universal choking sign, signs of poor air exchange and increased breathing difficulty, such as silent cough, cyanosis, or the inability to speak or breathe.
- Inability to provide positive pressure ventilation in a patient in respiratory or cardiac arrest after repositioning of the airway or placement of a BIAD.

Contraindications:
- None

Background:
FBAO occurs most commonly in adults while eating. In children FBAO may occur during eating or play. Most FBAO events are witnessed by bystanders and there is usually some intervention by bystanders prior to the arrival of EMS. In some instances however, FBAO may be unwitnessed and the patient may be found by EMS providers in unresponsive and in respiratory or cardiac arrest. In this instance, the presence of FBAO may only be recognized by the inability or difficulty in providing positive pressure ventilation.

Procedure:

Conscious Patient
1. Assess the patient to determine if the FBAO is complete (unable to speak) or partial (able to speak and or cough, displaying universal choking sign).
2. Do not interfere with a patient who has a mild/partial FBAO who is able to cough. Allow the patient to clear their airway by coughing and monitor closely.
3. For an **infant**, deliver 5 (five) back blows followed by five (5) chest thrusts repeatedly until the object is expelled or the patient becomes unresponsive. If the patient becomes unresponsive, manage as below in Unresponsive Patient.
4. For a **child**, perform sub diaphragmatic abdominal thrusts (Heimlich maneuver) until the object is expelled or the patient becomes unresponsive. If the patient becomes unresponsive, manage as below in Unresponsive Patient.
5. For an **adult**, perform sub diaphragmatic abdominal thrusts (Heimlich maneuver) until the object is expelled or the patient becomes unresponsive. Chest thrusts should be used in obese patients and in patients who are in the late stages of pregnancy. If the patient becomes unresponsive, manage as below in Unresponsive Patient.

Unresponsive Patient
1. Safely lower the patient to a hard surface.
2. Initiate CPR and resuscitative efforts following the age appropriate *Cardiac Arrest Protocol.*
3. Each time the airway is opened/accessed during CPR it should be observed for the presence of a foreign body. Do not perform blind finger sweeps.
4. A-C-P level providers should perform direct laryngoscopy to potentially identify and remove the foreign body utilizing Magill forceps and or suction*.
5. Provide airway management per the age appropriate *Airway Management Protocol*.

* Appropriately trained A-C level providers practicing within a service/system that does not include endotracheal intubation as part of the service/system scope of practice may still perform direct laryngoscopy and utilize Magill forceps for the purposes of foreign body airway obstruction removal.

E

Indication:
- Cardiac arrest, apnea or hypoventilation (A-C providers patients >8 yo with apnea or cardiac arrest only)
- Patient with loss of airway protective reflexes (ability to cough, swallow) or airway compromise (Paramedic only)
- Patient with inhalation or other injury with potential for evolving airway compromise (Paramedic only)

Contraindications:
- None

Background:
Endotracheal intubation (oral or nasal) provides the most definitive airway protection. While the risk to benefit ratio of oral intubation by a skilled provider to blindly inserted airway device (BIAD) insertion favors oral intubation, each case, including provider skill level and experience, must be evaluated individually. There may instances in which it may be preferable to manage a patient's airway with simple adjuncts and a bag-valve-mask device or a BIAD.

Procedure:
1. Perform an airway assessment to help determine potential for difficulty in performing bag-valve mask ventilation and/or intubation.
2. Prepare and check all equipment:
   - Functional suction
   - Bag-valve-mask attached to supplemental oxygen.
   - Stylet and/or ETTI available (if to be used).
   - Appropriate size ETT (cuff checked, tube and lubricate with water soluble lubricant)
   - Quantitative waveform capnography
3. Prepare, position, and pre-oxygenate the patient.
4. Manually open the patient's mouth and while holding the laryngoscope in the left hand, gently insert the blade (straight or curved) following the natural curvature of the tongue while displacing the tongue to the left.
5. Gently lift the laryngoscope upward and forward elevating the mandible without using the teeth as a fulcrum.
6. Direct the tip of the laryngoscope into the proper terminal location (vallecular space for the curved blade, over the epiglottis for the straight blade).
7. Visualize the laryngeal structures and pass the ETT through the vocal cords (the tube should be observed passing through the cords).
8. Remove the laryngoscope and then the stylet from the ETT. Hold on to the ETT when removing the stylet.
9. Inflate the ETT cuff with 5-10 ml of air.
10. Confirm proper placement of the ETT utilizing standard methods (presence of breath sounds, absence of gastric sounds) **and** quantitative waveform capnography (a colorimetric $EtCO_2$ device may be used for initial confirmation of placement if waveform capnography is not immediately available).
11. After confirmation of proper tube placement, ensure the tube is at the appropriate depth and secure the tube using a commercial device, tape, or plastic tubing. DO NOT secure the tube to the chin.
12. Required documentation includes ETT size, number of attempts,+/- success, tube depth (lip/dentation line), presence of bilateral breath sounds/absence of gastric sounds, presence of $EtCO_2$ waveform with quantitative $EtCO_2$ confirmation 6 breaths.
13. Endotracheal tube placement should be reconfirmed after every patient movement. Limit motion of the head, in relation to the torso after intubation. In particular, avoid flexion and extension. Such motion can dislodge the tube from the trachea, particularly in pediatric patients.

PEARLS:

- Colorimetric $EtCO_2$ devices are to be utilized for the initial confirmation of ETT placement only. Quantitative waveform capnography must be instituted as soon as possible and during transport.
- In cardiac arrest, chest compressions should not be interrupted placement of an advanced airway (BIAD or ETT).
- Performing laryngoscopy at arm's length allows for binocular vision and facilitates visualization of the glottic structures.
- After 6 ventilations, the $EtCO_2$ should be >10mmHg or comparable to pre-intubation values. If the $EtCO_2$ is <10 mmHg, check for adequate circulation and ventilation. If the $EtCO_2$ is <10 mmHg without physiologic explanation, remove the ETT and ventilate with a BVM, consider placement of a BIAD.
- The application of the Backward, Upward, and Rightward Pressure (BURP) Maneuver may facilitate visualization of the larynx. To perform the BURP maneuver, pressure is applied with the fingers over the thyroid cartilage and pressure is applied posteriorly, then cephalad (upwards) and, finally, laterally towards the patient's right.
- The BURP maneuver should not be confused with "cricoid pressure" or the Sellick maneuver (no longer recommended in emergency airway management) which were performed to reduce the risk of regurgitation of gastric contents during ETI.
- Cuffed endotracheal tube size (mm ID) for patient age 1-10 yo may be calculated by the age in years/4 + 3. Endotracheal tube size, however, is more reliably based on a child's body length. Length-based resuscitation tapes are helpful for children up to approximately 35 kg.
- The proper depth of insertion (lip line) of an endotracheal tube can be estimated by multiplying the tube diameter by 3 Example: 8.0 ETT X 3 = 24. It should be noted that this formula presumes an appropriate tube diameter.

Indication:
- Patient with respiratory failure or airway compromise in whom oral endotracheal intubation is not possible due to an intact gag reflex, trismus, angioedema, patient location/position or other condition.

Contraindications:
- Apnea.
- Suspected basilar skull or mid facial fractures are a relative contraindication (if the tube can be passed easily with good and continuous air movement, nasotracheal intubation can be safely performed).
- Patient on warfarin (Coumadin) or other anticoagulant/antiplatelet agents (relative contraindication).

Background:
There may be situations in which oral endotracheal intubation is not possible. Nasotracheal intubation provides an alternative means of performing endotracheal intubation. It should be noted that there may instances in which it may be preferable to manage a patient's airway with simple adjuncts and a bag-valve-mask device or a BIAD.

Procedure:
1. Prepare and check all equipment:
   - Functional suction
   - Bag-valve-mask attached to supplemental oxygen
   - Appropriate size endotracheal tube (ETT) [cuff checked, tube and lubricate with water soluble lubricant]
   - Quantitative waveform capnography
2. Identify the largest and least obstructed nare. Premedicate the nasal mucosa of the selected nare with 2% Lidocane Jelly and Neosynephrine.
3. Insert a nasopharyngeal airway into the pre-medicated nare to help dilate the nasal passage.
4. Prepare, position, and pre-oxygenate the patient.
5. Attach a quantitative capnographic sample line to the endotracheal tube adapter.
6. Remove the nasopharyngeal airway and begin to gently insert the ETT with the bevel facing toward the septum.
7. Continue to advance the ETT while listening for maximal air movement and the presence of a capnographic waveform.
8. At the point of max air movement (indicating proximity to the larynx), gently and evenly advance the ETT through the glottic opening during inspiration.
9. Upon entering the trachea, the patient may cough or gag. Do not remove the ETT, this is to be expected. The presence of a capnographic waveform should be noted at this time (the waveform will not be robust as the cuff of the ETT has not yet been inflated).
10. In the adult patient, the tube should be advanced to 28cm at the nare in a male and to 26cm in a female.
11. Inflate the cuff with 5-10 ml of air.
12. Confirm proper placement of the ETT utilizing standard methods (presence of breath sounds, absence of gastric sounds) **and** quantitative waveform capnography (colorimetric $EtCO_2$ devices are unreliable for confirming placement of nasally placed endotracheal tubes).
13. After confirmation of proper tube placement, ensure the tube is at the appropriate depth and secure the tube.
14. Required documentation includes ETT size, number of attempts,+/- success, tube depth (nare line), presence of bilateral breath sounds/absence of gastric sounds, presence of $EtCO_2$ waveform with quantitative $EtCO_2$ (preferred) or colorimetric $EtCO_2$ device confirmation 6 breaths.

15. Endotracheal tube placement should be reconfirmed after every patient movement. Limit motion of the head, in relation to the torso after intubation. In particular, avoid flexion and extension. Such motion can dislodge the tube from the trachea, particularly in pediatric patients.

P

PEARLS:
- Colorimetric EtCO$_2$ devices are to be utilized for the initial confirmation of ETT placement only. Quantitative waveform capnography must be instituted as soon as possible and during transport.
- Appropriate ETT size is usually one or two sizes down from the ideal oral tube size.
- Use of a nasal Rae or Endotrol© tube may facilitate this procedure.
- If, once beyond the nasopharynx, resistance to tube advancement is encountered, the tube may have become lodged in the pyriform sinus (tenting of the skin on either side of the thyroid cartilage may be observed). If this occurs, slightly withdraw the ETT and rotate it toward the midline and attempt to advance the tube again with the next inspiration.
- Patients who are on warfarin (Coumadin) or other anticoagulants/antiplatelet agents are at increased risk of bleeding with this procedure.

A
C
P

Indication:
- Known or anticipated difficult airway resulting from inability to visualize the vocal cords.
- To facilitate routine endotracheal intubation.

Contraindications:
- Patient requiring intubation with a < 6.0 endotracheal tube (ETT).

Background:
The use of an endotracheal tube introducer (ETTI) is designed to facilitate the passage of an endotracheal tube through the vocal cords when visualization of the glottic structures is limited to the arytenoid cartilages. The ETTI also helps facilitate placement of an endotracheal tube when supraglottic or laryngeal edema is present. The ETTI similar to a "gum-elastic bougie".

Procedure:
1. Lubricate the ETTI with water soluble lubricant.
2. Perform laryngoscopy utilizing a curved or straight blade.
3. When the laryngoscope blade is in place and exposing some of the laryngeal opening, advance the ETTI into the trachea. It may be possible to feel the tactile sensation of "clicking" as the tip of the introducer advances downward over the tracheal ring. The ETTI should be advanced until the thick black line on the proximal portion of the ETTI is aligned with the corner of the mouth.
4. Advance the ETT over the ETTI (it may be helpful to have an assistant slide the ETT over the ETTI).
5. Once the distal portion of the ETT enters the oropharynx and is approaching the glottic structures, the ETT should be rotated 90° <u>counterclockwise</u> allowing the ETT bevel to spread the arytenoid cartilages so that minimal force is used to pass the ETT.
6. The ETT should be advanced so the appropriate lip line marker is aligned at the lips.
7. While holding the ETT firmly in place, have an assistant remove the ETTI.
8. Remove the laryngoscope.
9. Inflate the ETT cuff
10. Confirm proper placement of the ETT utilizing standard methods (presence of breath sounds, absence of gastric sounds) **and** quantitative waveform capnography (a colorimetric $EtCO_2$ device may be used for initial confirmation of placement if waveform capnography is not immediately available).
11. Secure the ETT in place and provide standard care for an intubated patient.

PEARLS:
- The ETT may be preloaded over the ETTI and advanced at visualization of the glottic structures.
- Performing laryngoscopy at arm's length allows for binocular vision and facilitates visualization of the glottis structures.
- Rotating the ETT in a clockwise direction during passage will increase the chances of the ETT tip locking on the arytenoid cartilages prevent easy passage of the ETT into the trachea.

Indication:
- Primary means of airway management in cardiac arrest or in patients requiring ventilatory support when endotracheal intubation is unavailable or cannot be performed.
- Use as "rescue" airway in failed airway situation.

Contraindications:
- Patients who are conscious or who have an intact gag reflex.
- Patient outside of extremes of weight or height for airway size determination.
- Known esophageal disease/caustic ingestion (KING LT, LTS).
- Significant oral/neck trauma or hemorrhage.

Background:
There are a number of blindly inserted airway devices (BIAD) available for use. A BIAD may be used as a primary means of airway management or a "rescue" device in a failed airway Placement of a BIAD should not interrupt continuous chest compressions in cardiac arrest. BIAD do not afford the same level of airway protection as an endotracheal tube. This protocol addresses devices that are most commonly available to EMS providers. Services are encouraged to adopt one device for use throughout their system. Service Medical Directors are ultimately responsible for validating and documenting the competency of each member of the service in utilizing the service's device of choice.

Procedure:

King Airway Device (LT, LTS)
1. Select appropriate size device for the patient:

Size 2	12 - 25 kg/35-45" [90-115 cm] - King-LT(S) not available
Size 2.5	25 - 25 kg/41-51" [105-130 cm] - King LT(S) not available
Size 3	4 - 5' height [122-155 cm]
Size 4	5 -6' height [155-180 cm]
Size 5	> 6' height [>180 cm]

2. Test cuff inflation system for air leak.
3. Apply water soluble lubricant to the distal tip of the device.
4. Hold the device at the connector with the dominant hand. With the non-dominant hand hold the mouth open and apply chin lift, unless contraindicated by suspected cervical spine injury or patient position. Using a lateral approach, introduce the tip into the corner of the mouth.
5. Advance the tip behind the base of the tongue while rotating the tube back to the midline so the blue orientation line faces the patient's chin.
6. Without exerting excessive force, advance the tube until the base of the connector is aligned with the teeth or gums.
7. Inflate the cuff to the appropriate volume (or to 60 cmH$_2$O or to "just seal" volume) utilizing the supplied syringe.
8. If necessary, additional volume may be added to the cuff to maximize seal of the airway.
9. Confirm proper **placement of the LTS** utilizing standard methods (presence of breath sounds, absence of gastric sounds) **and** by use of a colorimetric EtCO$_2$ detection device, quantitative waveform capnography may be used if available.

I-gel Airway
1. Select appropriate size device for the patient:

Size 1	2 - 5 kg	(neonate)
Size 1.5	5 - 12 kg	(infant)
Size 2	10 - 25 kg	(pediatric, small)
Size 2.5	25 - 35 kg	(pediatric, large)
Size 3	30 - 60 kg	(adult, small)
Size 4	50 - 90 kg	(adult, medium)
Size 5	90 kg	(adult, large)

E

E

2. Open the I-gel package, and on a flat surface take out the protective cradle containing the device.
3. Remove dentures or removable dental plates from the mouth prior to insertion.
4. Remove the I-gel and transfer it to the palm of the same hand that is holding the protective cradle, supporting the device between the thumb and index finger.
5. Place a small bolus of a water soluble lubricant onto the middle of the smooth surface of the protective cradle in preparation for lubrication.
6. Grasp the I-gel with the opposite (free) hand along the integral bite block and lubricate the back, sides and front of the cuff with a thin layer of lubricant.
7. Place the i-gel back into the protective cradle in preparation for insertion.
8. Unless there is concern for cervical spine injury, place the patient in the in the "sniffing" position (head extended and neck flexed).
9. Grasp the lubricated i-gel firmly along the integral bite block. The device should be held so that the i-gel cuff outlet is facing towards the chin of the patient.
10. While gently pressing the chin down, introduce the leading soft tip into the mouth of the patient in a direction towards the hard palate.
11. Glide the device downward and backward along the hard palate with a continuous but gentle push until a definitive resistance is felt.
12. The tip of the airway should be located into the upper esophageal opening and the cuff should be located against the laryngeal framework. The incisors should be resting on the integral bite-block.
13. The i-gel should be taped down from maxilla to maxilla and secured with the airway support strap provided.
14. Confirm proper placement of the device utilizing standard methods (presence of breath sounds **and** colorimetric EtCO2 detection device **or** waveform capnography (preferred).
15. Once proper position is confirmed, it is recommended that a provider stabilize the device with a free hand, particularly during patient movement.

Air-Q Laryngeal Airway

1. Select appropriate size device for the patient:

Size 0.5	< 4 kg
Size 1.0	4-7 kg
Size 1.5	7-10 kg
Size 2.0	17-30 kg
Size 2.5	30 - 50 kg
Size 3.5	50 - 70 kg
Size 4.5	70 - 100 kg

2. Ensure the inflation valve is open to air (either leave the supplied tag in place or attach an empty syringe barrel to the inflation valve.
3. Lubricate the external surface of the device including the mask and ridges with water soluble lubricant.
4. Open the patient's mouth and elevate the tongue (this can be accomplished by performing a mandibular lift [recommended] or by inserting a tongue blade at the base of the tongue and lifting).
5. Place the front portion of the air-Q mask between the base of the tongue and the soft palate at a slight angle.
6. Insert the device into position within the pharynx by gently applying inward and downward pressure (minimal manipulation may be required to facilitate passage around the corner and into the upper pharynx). Continue to advance until fixed resistance to forward movement is felt.
7. Tape the device in place and inflate the cuff with the appropriate volume of air for the device selected.
8. Confirm proper placement of the device utilizing standard methods (presence of breath sounds and colorimetric $EtCO_2$ detection device or waveform capnography (preferred).

E

Laryngeal Mask Airway (LMA)™

First and second generation LMAs are available in a number of variations. These variations include, but are not limited to the LMA Supreme™, LMA Unique™ and the LMA Fastrach™. Each device has a unique recommended insertion technique. Providers using any variation of the LMA must be familiar with the insertion technique specific to the device being used. Below is the general insertion technique for some of the first generation devices:

1. Select the appropriate size device for the patient.
2. Confirm device cuff integrity and then deflate the cuff completely.
3. Apply water soluble lubricant only to the posterior surface of the device.
4. Unless cervical spine injury is of concerns, position the head in a slight sniffing position (neck flexed and head extended). This position should be maintained by pushing the head from behind with the non-dominant hand.
5. While holding the device like a pen with the index finger placed anteriorly at the junction of the cuff and tube, insert the device into the oropharynx and press the tip of the device against the hard palate and verify that the tip is against the hard palate and not folded over.
6. Using the index finger, insert the mask backwards while still maintaining pressure against the hard palate.
7. Insert the index finger completely into the hypopharynx until resistance is met.
8. Stabilize the device by grasping the LMA tube with the non-dominant hand and remove the index finger.
9. Ensure that the black marker line on the airway tube is oriented anteriorly toward the upper lip.
10. Without holding the tube, inflate the cuff with the appropriate volume of air for the selected device (the device should position itself as the cuff is inflated).
11. Confirm proper placement of the device utilizing standard methods (presence of breath sounds **and** colorimetric $EtCO_2$ detection device **or** waveform capnography (preferred).
12. Stabilize the device in using tape.

PEARLS:

- In cardiac arrest, chest compressions should not be interrupted placement of an advanced airway (BIAD or ETT).

Indication:
- Failed airway (can't intubate, can't ventilate, can't oxygenate).

Contraindications:
- None in the failed airway situation.

Background:
Cricothyrotomy is an infrequently preformed, but potentially lifesaving procedure. There are several methods of performing cricothyrtomy. Conventional surgical cricothyrotomy entails surgically providing an opening in the cricothyroid membrane for placement of an artificial airway. A variation of this technique, the endotracheal tube intoducer [ETTI] ("Bougie") assisted technique, utilizes an ETTI in a method similar to the Seldinger technique to introduce an ETT through the cricothyroid membrane via a surgical incision. The percutaneous dilational technique introduces an airway catheter into the cricothyroid membrane space over a guidewire and dilator (Seldinger technique). Finally, needle cricothyrotomy utilizes a catheter inserted over a needle into the cricothyroid space to provide transtracheal ventilation. <u>Needle cricothyrotomy is reserved for patients < 12 yo and is considered a temporary means of airway access</u>. The ability to provide adequate ventilation via a needle cricothyrotomy is limited.

Procedure:

Conventional Surgical Technique
1. Position the patient in a supine position with the head in a neutral position.
2. Identify and palpate the thyroid cartilage, cricoid membrane, and cricoid cartilage.
3. Prepare the area over the cricothyroid membrane with alcohol or chlorohexidine.
4. Stabilize the thyroid cartilage with thumb and index finger with the non-dominant hand.
5. Identify the cricothyroid membrane with the index finger of the non-dominant hand.
6. Utilizing a scalpel with a #11 blade, poke the membrane to make an incision across the lower 1/3 of the cricothyroid membrane.
7. Insert the left finger into the cricothyroid space to re-identify the membrane and the incision.
8. Insert a tracheal hook (if available) into the incision (the hook should be held at a right angle to the patient's neck and directed into the incision using the index finger. Once the hook is located in the airway, rotate the hook in a cephalad direction and hold it at a 45° angle.
9. Insert a Trousseau dilator or curved hemostat into the incision and gently spread the incision to increase the vertical diameter of the membrane incision.
10. Insert a #4.0 cuffed tracheostomy tube or a # 6.0 endotracheal tube.
11. Remove the obturator if a tracheostomy tube was used.
12. Inflate the device cuff with appropriate volume of air.
13. Confirm proper placement of the airway device utilizing standard methods (presence of breath sounds, absence of gastric sounds) and quantitative waveform capnography (a colorimetric $EtCO_2$ device may be used for initial confirmation of placement if waveform capnography is not immediately available).
14. Secure the airway device in place and provide standard care for the intubated patient.

P

"Bouige" Assisted Cricothyrotomy Technique
1. Position the patient in a supine position with the head in a neutral position.
2. Identify and palpate the thyroid cartilage, cricoid membrane, and cricoid cartilage.
3. Prepare the area over the cricothyroid membrane with alcohol or chlorohexidine.
4. Standing on the patients left side, stabilize the larynx with the thumb and index finger of your left hand and identify the cricothyroid membrane.
5. Using a scalpel with a #20 blade make a stabbing incision through the skin and cricothyroid membrane.
6. Place a tracheal hook at the inferior margin of the incision and gently pull up on the trachea.
7. Insert an endotracheal tube introducer (ETTI, "Bouigie") through the incision in a caudal direction.
8. Place a 6.0 cuffed endotracheal tube over the ETTI and into the trachea.
9. Remove the ETTI and inflate the ETT cuff with 5-10 ml of air. .
10. Confirm proper placement of the airway device utilizing standard methods (presence of breath sounds, absence of gastric sounds) and quantitative waveform capnography (a colorimetric EtCO2 device may be used for initial confirmation of placement if waveform capnography is not immediately available).
14. Secure the airway device in place and provide standard care for the intubated patient.

Percutaneous Dilational Cricothyrotomy
1. Position the patient in a supine position with the head in a neutral position.
2. Identify and palpate the thyroid cartilage, cricoid membrane, and cricoid cartilage.
3. Prepare the area over the cricothyroid membrane with alcohol or chlorohexidine
4. Utilizing a scalpel with a #15 blade, make a 2-3 cm midline vertical incision over the cricothyroid membrane.
5. With a syringe attached, advance the introducer needle and catheter through the incision and into the airway at a 45 degree angle to the frontal plane in the midline in a caudal direction.
6. Aspirate with the attached syringe to confirm proper placement of the needle and introducer by the aspiration of free air.
7. Once proper placement has been confirmed, remove the needle while leaving the introducer catheter in place.
8. Advance the airway access assembly (with dilator inserted) over the guidewire until the proximal stiff end of the guidewire is completely through and visible at the handle end of the dilator.
9. Remove the guidewire and dilator simultaneously.
10. Inflate the airway cuff with 5-10 ml of air.
11. Confirm proper placement of the airway device utilizing standard methods (presence of breath sounds, absence of gastric sounds) and quantitative waveform capnography (a colorimetric EtCO2 device may be used for initial confirmation of placement if waveform capnography is not immediately available).
12. Secure the airway device in place and provide standard care for the intubated patient.

P

Control-Cric™

1. Position the patient supine and identify the cricothyroid membrane.
2. Stabilize the larynx with thumb and middle finger with the non-dominant hand.
3. Utilizing the Cric-Knife, incise the skin making a vertical incision from the mid-thyroid cartilage to the cricoid cartilage (about 2 finger breadths in length). A longer incision may be needed if the patient has a thick neck. If landmarks are clearly visible, a horizontal incision may be used.
4. After palpating the cricothyroid membrane, turn the Cric-Knife to a horizontal position over the cricothyroid membrane.
5. Push the blade downward, perpendicular to the trachea, until the blade is fully inserted and the airway is entered.
6. While maintaining a downward force, slide the tracheal hook down the handle with your thumb until the hook is felt to enter the trachea, and disengages from the handle.
7. Grab the hook with the non-dominant hand, lifting up on the thyroid cartilage.
8. Insert the Cric-Key through the incision. Placement can be confirmed by moving the device along the anterior wall of the trachea to feel for the tracheal rings. Tenting of the skin, difficulty advancing the Cric-Key, or lack of tactile feedback from the tracheal rings suggests incorrect placement.
9. Once placement has been confirmed, advance the Cric-Key to the flange. Stabilize the Cric-Key tube and pivot the tracheal hook toward the patient's shoulder to remove from the airway.
10. While stabilizing the Cric-Key tube, remove the Cric-Key introducer. Inflate the cuff until resistance is met.
11. Confirm proper placement of the airway device utilizing standard methods (presence of breath sounds, absence of gastric sounds) and quantitative waveform capnography (a colorimetric $EtCO_2$ device may be used for initial confirmation of placement if waveform capnography is not immediately available).
12. Secure the device with the stabilizing strap and airway device in place and provide standard care for the intubated patient.

Needle Cricothyrotomy

1. Position the patient in a supine position with the head in a neutral position.
2. Identify and palpate the thyroid cartilage, cricoid membrane, and cricoid cartilage.
3. Prepare the area over the cricothyroid membrane with alcohol or chlorohexidine.
4. Attach a 10 ml syringe to a transtracheal catheter (preferred) or a 14-16g catheter over the needle device.
5. Advance a transtracheal catheter/or catheter needle over the device in a caudal direction through the cricothyroid membrane and into the trachea. Be cautious to avoid penetrating the posterior trachea.
6. Remove the needle from the transtracheal airway catheter/catheter over the needle device.
7. Attach a high pressure (50 psi) ventilation device (patients < 5 yo should be ventilated with a bag-valve-mask device) to the proximal end of the airway catheter and ventilate the patient while observing for chest rise and auscultating breath sounds.
8. Continue ventilating while manually stabilizing the airway catheter, provide ventilations at a rate of 12-20 ventilations per minute with an I:E ratio of 1:2 or 1:3 seconds. For patients < 5yo, ventilate until chest rise is observed. For patients 5-12 yo, ventilate at inspiratory pressures of 30-50 mmHg.

P

PEARLS:

- Being a low frequency procedure, providers should maintain familiarization with related anatomy and the technical aspects of performing cricothyrotomy on at least a yearly basis. This can be achieved by the use of cadaveric or other anatomic models and skills trainers.
- The most common errors in performing cricothyrotomy are related to inaccurate landmarking and therefore, inaccurate incision.
- The SHORT mnemonic may be used for recalling factors potentially associated with difficult cricothyrotomy:

    Surgery (history of neck surgery, presence of a surgical scar)
    Hematoma
    Obesity
    Radiation (history or evidence of XRT)
    Trauma (direct laryngeal trauma with displaced landmarks)
- Incision is made through the inferior edge (lower 1/3) of the cricothyroid membrane due to the relatively cephalad location artery and vein, which run transversely near the top of the membrane.
- In children, the cricothyroid membrane is disproportionately smaller because of greater overlap of the thyroid cartilage over the cricoid cartilage. For this reason, surgical cricothyrotomy is not recommended in patients 12 years of age or younger.
- Needle cricothyrotomy is generally reserved for patients <12 yo (not generally a procedure to be performed on the adult patient).
- Use of a nonkinkable wire-coiled transtracheal catheter jet ventilation catheter is preferred over a catheter over the needle intravenous catheter for needle cricothyrotomy.
- A 35mm adapter from a 3.0 ETT will fit the Luer lock connector of a transtracheal catheter or that of a catheter over the needle device.

Indication:
- Patient with asthma, reactive airway disease, bronchospasm.

Contraindications:
- Patient unable to cooperate with test.

Background:
The peak expiratory flow (PEF), also called peak expiratory flow rate (PEFR) reflects a patient's maximum speed of expiration. In patients with asthma, establishing a PEFR baseline allows the provider to assess the patient's response to bronchodilator therapy. When used in the management of acute asthma, measurement of the PEFR is taken pre and post bronchodilator therapy. The PEFR is usually measured three times in succession and the highest value of the three measurements is documented. Normal PEFR varies with height. Many asthmatics routinely measure their PEFR and can provide the EMS provider with their normal PEFR. Measurement of PEFR should not delay treatment in the patient with severe acute asthma.

Procedure:
1. Position patient in an upright position (45-90° angle).
2. Hold the PEFR meter upright with the mouthpiece several from the patient's mouth.
3. Instruct the patient to inhale deeply as possible and place his/her mouth firmly around the mouthpiece of the PEFR meter making sure that a tight seal is created by the patient's lips.
4. Instruct the patient to exhale as hard as possible.
5. Note the PEFR measurement.
6. Repeat steps above two additional times.
7. Document the highest value of the three measurements.

P

PEARLS:
- Normal PEFR can be approximated using the following formula: PEFR (L/min) = [Height (cm) - 80] X 5.

Indication:
- Initial confirmation of proper placement of an advanced airway (endotracheal tube [ETT] or blindly inserted airway device [BIAD]).

Contraindications:
- None.

Background:
A colorimetric end tidal carbon dioxide detection device may be used as an adjunct in the initial confirmation of advanced airway (ETT or BIAD) placement. The colorimetric device is similar to litmus (pH) paper and is purple at baseline, turning lighter (yellow) in the presence of CO_2. The presence of CO_2 generally indicates proper (tracheal, not esophageal) placement of an advanced airway. However, a detector's indication may be incorrect if there is no exhaled CO_2 from the lungs (e.g. in situations of circulatory arrest with no tissue perfusion), or if there is CO_2 from the stomach (e.g. when carbonated beverages have been ingested). The presence of gastric acid, as from vomiting, may also result in color change despite esophageal ETT positioning (this "false-positive" color change will usually not vary with respiratory cycle). There are several manufactures of these devices. Reference in this protocol to patient weight based patient selection (adult v. pediatric) and color change correlations to EtCO2 levels is specific to the Nellcor Easy Cap II and Pedi-Cap CO2 detection devices manufactured by Covidien. Providers should reference the specifications of their service specific device of choice.

A

C

P

Procedure:
1. Select appropriate detector depending on patient size and weight (Easy-Cap for patients 15 kg, Pedi-Cap for patients ≥1-15 kg).
2. Attach the detector device to the airway device (ETT or BIAD).
3. Attach a bag-valve-device to the detector device.
4. Deliver 6 ventilations of moderate tidal volume.
5. After the delivery of 6 ventilations as above, observe the color in the indicator window on full-end expiration. If CO_2 is detected, the purple indicator will change from PURPLE to TAN (0.5-2.0%) or YELLOW (2-5%). The color change should be phasic with the ventilator cycle. This cycling of the colorimetric device in the YELLOW range is an extremely reliable indicator of appropriate ETT placement.
6. If the results are not conclusive, and the correct anatomic location of the advanced airway cannot be confirmed with certainty by other means, the advanced airway should be removed.

PEARLS:
- Colorimetric EtCO2 devices are to be utilized for the initial confirmation of ETT placement only. Quantitative waveform capnography must be instituted as soon as possible and during transport.
- Colorimetric EtCO2 detection devices are not reliable for confirming placement of nasally placed endotracheal tubes
- Contamination of the detector with gastric contents, medication, or mucous may increase resistance, alter color change, and affect ventilation. If the device becomes contaminated, it should be discarded and replaced.
- The detector device has a useful life of about two hours and may be left in place for that period of time.
- An adult detector device may be used for a pediatric patient for initial airway placement confirmation, but due concerns for increased dead space, it must not be left in place.
- Due to the potential for increased airway resistance, do not use the pediatric device on patients ≥15 kg. Further, if an adult device is used in a pediatric patients, low tidal volumes may prevent color change even if ETT placement is correct (tracheal).
- The colorimetric device is not a replacement for waveform capnography.

A C P

Indication:
- Confirmation and monitoring of airway placement (ETT, BIAD, cricothyrotomy).
- Monitoring of ventilatory status in patients receiving sedation and analgesia.
- Cardiac arrest, asthma, reactive airway disease, respiratory distress, suspected DKA, sepsis or pulmonary embolus.

Contraindications:
- None.

Background:
Quantitative waveform capnography is the continuous measurement of carbon dioxide (CO_2), specifically end-tidal CO_2. The capnograph provides information not only regarding pulmonary ventilatory function, but also indirect information regarding cardiac function and perfusion. In addition to confirming and monitoring airway placement, quantitative waveform capnography has many applications in the critically ill patient. Waveform capnography allows providers to monitor CPR quality, optimize chest compressions, detect return of spontaneous circulation during chest compressions, and assess cardiac output in patients with pulseless electrical activity. EtCO2 monitoring can be used to guide ventilation parameters and assess the severity of respiratory distress and ventilatory fatigue (CO_2 retention) in a number of pathological conditions and in patients receiving sedation and or narcotic analgesia. It is also useful in assessing the degree of circulatory failure in shock from any cause. Capnography can also be used as part of screening patients for DKA or sepsis. The EtCO2 waveform provides information related to airway obstruction and bronchospasm.

Procedure:
1. Select age appropriate sampling line/sensor.
2. Attach sample line to the EtCO2 monitoring device, verify the EtCO2 display is on and functioning.
3. Attach the sampling line/sensor to the ETT, BIAD or oxygen delivery device in the spontaneously breathing patient without an artificial airway.
4. After 6 breaths, note the EtCO2 level and waveform characteristics.
5. In all patients with spontaneous circulation, an EtCO2 of 20mmHg is anticipated (35-45 mmHg in patients with normal cardiac and pulmonary function). In patients undergoing CPR, if the EtCO2 is <15 mmHg, the quality (rate, depth, recoil) of external chest compressions should be assessed. In the post-resuscitation patient, no effort should be made to lower ETCO2 by modification of the ventilation rate.
6. Any loss of CO_2 detection or waveform must be addressed immediately. Consider the following:
   o Apnea or loss of airway (tube dislodgement, esophageal placement, obstruction).
   o Circulatory collapse (cardiac arrest, exsanguination, massive pulmonary embolism).
   o Equipment failure (disconnection from ventilation device, equipment malfunction).
7. Document the use of capnography. Serial EtCO2 levels should be documented with each set of vital signs.

PEARLS:
- The three physiologic factors affecting the EtCO2 are metabolism, ventilation and cardiac output. If any two these factors remain constant, any change in the EtCO2 can be attributed to the third.
- In cardiac arrest, the EtCO2 is reflective of cardiac output generated by external chest compressions.

Indication:
- Patient with obstruction of the airway secondary to secretions, blood, or other substances.

Contraindications:
- None

Background:
Patients with loss of airway protection (inability to cough or swallow) may require suctioning for the relief of airway obstruction secondary to secretions, blood, or other substances.

Procedure:
1.   Ensure suction apparatus is in proper working order.
2.   Attach desired suction tip to suction line.
3.   Pre-oxygenate the patient as much as possible.
4.   Utilize appropriate body substance isolation precautions.
5.   Examine the oropharynx and remove any potential foreign objects or materials which may occlude the airway if dislodged by the suction device.
6.   Utilize the suction device to remove any secretions, blood or other substance. The alert and cooperative patient may assist with this procedure.
7.   Reattach ventilator device (e.g. bag-valve-mask) and ventilate or assist the patient as indicated.

E

Indication:
- Patient with an endotracheal or tracheostomy tube in place requiring deep suctioning for relief of airway obstruction secondary to secretions, blood, or any other substance.

Contraindications:
- None

Background:
Patients with an artificial airway in place (endotracheal tube, tracheostomy tube, cricothyrotomy) may require deep endotracheal suctioning for the relief of airway obstruction secondary to secretions, blood, or other substances. Endotracheal suctioning requires use of aseptic technique.

Procedure:
1. Ensure suction apparatus is in proper working order.
2. Pre-oxygenate the patient as much is possible (use FiO2 of 1.0 for a minimum of 30 seconds if able).
3. Utilize appropriate body substance isolation precautions.
4. Select the appropriate size flexible suction catheter for the device to be suctioned.
5. Determine the appropriate depth for insertion of the suction catheter by using the distance from the suprasternal notch to the proximal of the airway into which the catheter will be placed (judgement must be used regarding the depth of suctioning for cricothyrotomy and tracheostomy tubes).
6. Set the suction to -100 to -120 mmHg for the adult patient (-80 to -100 mmHg for children and -60 to -80 mmHg for infants).
7. While maintaining aseptic technique, open the suction catheter package and attach the connecting tube to the suction catheter.
8. Detach the patient from the BVM or ventilator.
9. Insert the flexible suction catheter as determined above.
10. Once the predetermined depth has been reached with the suction catheter, occlude the thumb port and slowly remove the suction catheter.
11. Reattach the ventilator device (BVM, ventilator) and ventilate the patient.
12. Suction the oropharynx as required.

P

PEARLS:
- Suction catheter size can be determined by multiplying the endotracheal tube's inner diameter (ID) by 2. Use the next smallest size catheter. Example: 6.0 mm endotracheal tube: 2 x 6 = 12; next smallest size catheter is 10 French.
- When obstruction/resistance is met, retract the catheter slightly prior to applying suction. Obstruction suggest the catheter has been advanced to the carina or a main bronchus. If suction is applied at this time, mucosal injury may occur.

Indication:
- Patient with tracheostomy with urgent or emergent indication to change or replace a tracheostomy tube, such as obstruction that will not clear with suction, dislodgement, or the inability to maintain oxygenation and or ventilation without obvious explanation.

Contraindications:
- None

Background:
Patients who have undergone tracheostomy may maintain a natural stoma or may have a tracheostomy tube in place. There are multiple types of tracheostomy tubes. Common types include cuffed with a disposable inner cannula (DIC), cuffed with a reusable inner cannula, cuffless with a DIC, cufflless with a reusable inner cannula, cuffed fenestrated, cuffless fenestrated, and metal tracheostomy tubes. A tracheostomy is not mature until after two weeks. Great caution should be exercised in attempting to change an immature tracheostomy site.

Procedure:
1. Have all equipment prepared for standard airway management, including equipment for orotracheal intubation and failed airway.
2. Have airway device (endotracheal tube or tracheostomy tube) of the same diameter of the device that is in situ as well as one that is 0.5 size smaller.
3. Lubricate the replacement tube(s) and check cuff integrity.
4. If the patient is receiving mechanical ventilation, detach the ventilator and ventilate the patient with a bag-valve-mask device to preoxygenate the patient as much as possible.
5. Once all equipment is in place, remove any securing devices from the in situ device, including sutures and or supporting dressings.
6. If applicable, deflate the cuff of the in situ device. If unable to aspirate air from the cuff with a syringe, cut the pilot balloon off to allow the cuff to lose pressure.
7. Remove the in situ device.
8. Insert the replacement tube. Confirm placement via standard airway placement confirmation techniques, except use of an esophageal detection device (which is ineffective for surgical airways).
9. If there is any difficulty placing the tube, re-attempt utilizing the smaller tube.
10. If difficulty is still encountered, use standard airway procedures such as oral bag-valve-mask ventilation or oral endotracheal intubation.
11. Document procedure, confirmation of proper placement, patient response and any complications.

P

PEARLS:
- More difficulty with changing the tube should be anticipated for sites that are immature (< 2 weeks old). Great caution must exercised in attempts to change immature tracheostomy sites.

Indication:
- Patient requiring medication administration via nebulized route.

Contraindications:
- None

Background:
Certain medications may be administered via nebulization. Administration of medication via this route is advantageous in some clinical situations as the medication is deposited directly into the respiratory tract and thus higher drug concentrations can be achieved in the bronchial tree and pulmonary bed with fewer adverse effects than when the systemic route is utilized.

Procedure:
Non-Intubated Patient

1. Explain procedure to patient if applicable.
2. Determine baseline pulse rate, SpO_2, PEFR (if to be measured), and auscultate lung sounds.
3. Confirm desired medication and absence of known allergy to selected medication.
4. Assemble small volume nebulizer (SVN). If a face mask is being utilized, attach the female fitting on the bottom of the mask directly to the male adapter on the medication port.
5. Instill the desired dose of medication, and if applicable, the appropriate amount of saline into the reservoir well of the SVN.
6. Attach oxygen supply tubing to the SVN.
7. Set oxygen flow rate per device specifications.
8. Ensure that medication is flowing prior to giving mouthpiece to the patient or applying the mask on the patient.
9. Place the mouthpiece in the patient's mouth or position the face mask on the patient, instructing them to breathe normally and take a deep breath every 3-5 inhalations.
10. Discontinue treatment when solution is depleted.
11. Place patient on supplemental oxygen if indicated.
12. Reassess pulse rate, SpO_2, PEFR (if to be measured), and auscultate lung sounds.
13. Document medication administration including dose and time as well as any patient response.

Intubated Patient

Follow steps 1-11 above for the non-intubated patient.
1. Attach the free end of the 6"corrugated tubing to the non-rebreathing exhalation port of the bag-valve-mask device.
2. Ensure the suctioning port (if present) on the 90 degree adapter is closed.
3. Hand ventilate the patient.

Patient Receiving CPAP

See *Continuous Positive Airway Pressure (CPAP) Protocol*

E

PEARLS:
- Medications that can be administered (per protocol) include albuterol, ipratropium bromide, epinephrine, levalbuterol, and calcium gluconate.

A
C
P

Indication:
- Patient requiring medication administration via a metered-dose inhaler.

Contraindications:
- None

Background:
Certain medications may be administered via a metered-dose inhaler (MDI). Similar in nature to medication administration via a nebulizer, this route is advantages in some clinical situations as the medication is deposited directly into the respiratory tract and thus higher drug concentrations can be achieved in the bronchial tree and pulmonary bed with fewer adverse effects than when the systemic route is utilized.

Procedure:
Non-intubated Patient
1. Explain procedure to patient if applicable.
2. Determine baseline pulse rate, SpO2, PEFR (if to be measured), and auscultate lung sounds.
3. Confirm desired medication and absence of known allergy to selected medication.
4. Assemble the MDI (if required) and warm the MDI to hand or body temperature.
5. Hold the MDI upright and shake the canister vigorously to mix the contents.
6. Attach a spacer if one is to be used.
7. If able, position the patient in an upright position (45-90°).
8. Place the MDI approximately 4 cm from the patient's open mouth. If a spacer is being utilized, place the spacer mouthpiece between the patient's teeth and instruct the patient to close their lips around it.
9. Instruct the patient to exhale completely and then begin inhaling slowly while the MDI is actuated. The patient should continue to inhale to total lung capacity (TLC). Once TLC is reached, the patient should be instructed to hold their breath for ten seconds.
10. Place patient on supplemental oxygen if indicated.
11. Reassess pulse rate, SpO2, PEFR (if to be measured), and auscultate lung sounds.
12. Document medication administration including dose and time as well as any patient response.

PEARLS:
- The use of a spacer makes the use of a MDI easier, particularly in children.
- Spacers make it easier for medication to reach the lungs and results in less medication being deposited in the upper airway.

Indication:
- Patient with potential or suspected hypoxemia.

Contraindications:
- None

Background:
An arterial blood-oxygen saturation reading indicates the percentage of hemoglobin (Hb) molecules in the arterial blood which are saturated oxygen (O2).   When measured by the use of pulse oximetry, this measurement is referred to as the SpO2. An oximeter is not able to differentiate between Hb which is saturated by oxygen (HbO2) vs Hb which is saturated by carbon monoxide (HbCO) or methemoglobin (METHb). A co-oximeter must be used for this purpose. In general, a normal oxygen saturation is 97-99%. Measurements below 92-94% suggest respiratory compromise of a hypoxic nature. Supplemental oxygen administration is usually not indicated if the SpO2 is ≥ 94%.

E

Procedure:
1. Select appropriate site for application of the oximetric sensor. Utilize a site recommended by the device manufacturer. Most sensors work on the fingers, toes or ear. If non-invasive blood pressure monitoring (NIBP) is also being utilized, consider placement of the sensor on an extremity other than the one being utilized for NIBP.
2. Allow oximeter to register the saturation level.
3. Verify pulse rate displayed by the oximeter with the patient's pulse. If waveform oximetry is being utilized, assess the quality of the atrial waveform displayed to help validate the saturation level registered.
4. If performing a one time ("spot") measurement, the probe/sensor should be left in place for a few minutes as oxygen saturation can vary.
5. Document the SpO2 every time vital signs are documented and in response to therapy to correct hypoxemia.

PEARLS:
- If a probe/sensor is too large or small, the light emitting diode (LED) and the light detector might not line up and result in a false reading.
- Patients with anemia have a reduced level of functional Hb, however this limited amount of functional Hb may be well saturated with oxygen, so the patient may have a normal SpO2 in spite of having a reduced oxygen carrying capacity and potential hypoxia.
- In patients with poor perfusion/vasoconstriction (shock) it may be difficult for the oximeter to pick up an adequate signal due to reduced pulsatile blood flow. Similarly, this may occur in the hypothermic patient.
- Irregular cardiac rhythms or tachycardia may reduce the reliability of oximetric readings.
- Some nail polish or prosthetic fingernails may interfere with the accuracy of oximetry. If possible, select a digit with an unpolished nail, rotate the probe 90 degrees so the light is shining through the finger sideways, consider another site, or utilize acetone to remove the nail polish.
- Sometimes external light sources may cause inaccurate readings. If this is suspected, try covering the site with opaque material.
- The presence of an arterial oximetric waveform may be useful in verifying the presence of mechanical capture during transcutaneous pacing.
- The preferred probe/sensor application site for newborns immediately after birth is the right hand. SpO2 values measured on the right hand (pre-ductal) are more representative of brain oxygenation.

Indication:
- Patient requiring ventilator support.

Contraindications:
- Agonal breathing/respiratory arrest (BIPAP)
- Inability to protect airway/copious secretions or vomiting (BIPAP)
- Decreased level of consciousness (BIPAP)
- Facial surgery/trauma precluding mask seal (BIPAP)

E

1. Non-paramedic level EMS providers may transport patients who are chronically mechanically ventilated on a home/personal mechanical ventilatory device.
2. The patient must be accompanied by an individual (family member or healthcare provider) who normally provides ventilator care for the patient and is familiar with the device and the patients ventilatory needs, can troubleshooting the device, and is able to provide deep tracheal suctioning, if required.
3. The patient's ventilator must be used for the transport.
4. Non-paramedic level providers may not adjust or otherwise manipulate ventilator settings.

Background:
Paramedic level providers may institute mechanical ventilation in the adult patient after placement of an advanced airway in the field and may maintain mechanical ventilation in patients undergoing interfacility transport when mechanical ventilation (invasive and non-invasive) has been initiated by the sending facility. Maintenance of mechanical ventilation is limited to patients being ventilated with a conventional mode of ventilation. Services utilizing mechanical ventilators are responsible for validating the competency of individual providers in the use of the service specific ventilator(s) and in the management of the mechanically ventilated patient. Documentation of individual competency must be maintained by the service.

Procedure:

General

P

1. All ventilators must be able to meet the demands of the patient's condition, taking into consideration all settings and features described or stipulated by the sending facility.
2. **If providers (non CCT paramedic) encounter a patient receiving a non-conventional mode of ventilation (e.g. adaptive support ventilation, airway pressure release ventilation, dual control modes, proportional assist ventilation, volume-assured pressure support, volume support etc.) with other complex ventilator needs, or a diagnosis of acute respiratory distress syndrome (ARDS), arrangements should be made to have the patient transported by a critical care specialty transport team or to have a critical care practitioner familiar and experienced with the specific mode of ventilation/needs accompany the patient.**
3. Unless the transfer is time sensitive in nature (e.g., STEMI, aortic dissection, acute CVA, unstable trauma, etc.), patients should be on the transport ventilator for 20 minutes prior to departure from the sending facility.
4. If there is question of ventilator function or troubleshooting of unresolved alarms is required, the patient should be disconnected from the ventilator and manually ventilated until the problem/alarm is resolved.
5. An appropriate size bag-valve-mask device, oropharyngeal airway, and 10 ml syringe must be with the patient and immediately available at all times.

P

6.   The following should be documented (as applicable) in the PCR for all mechanically ventilated patients: mode of ventilation, set rate/total rate, FiO_2, positive inspiratory pressure (PIP), pressure support, positive end expiratory pressure (PEEP) and the plateau pressure (if measured).
7.   Ensure/calculate gas supply required for duration of transport and emergency reserve.
8.   Procedural steps below are as applicable to specific to the ventilator and mode be used.

Initiating Mechanical Ventilation after Placement of an Advanced Airway by EM (assumes use of volume controlled ventilation and assist control [AC] mode):

1.   Turn the ventilator on and allow it to self- test/cycle as required for the ventilator being utilized.
2.   Select AC mode.
3.   Set the desired tidal volume: 8 ml/kg (larger volumes may lead to ventilator induced lung injury). The predicted body weight (PBW) should be utilized for this calculation: Males = 50 + 2.3 [height (inches) - 60], Females = 45.5 + 2.3 [height (inches) -60]. See included reference chart.
4.   Set the desired rate (generally 10-12 bpm).
5.   Set the desired FiO_2 (1.0 may be used initially, but it should be decreased as soon as possible to the lowest level that maintains the SpO2 of ≥ 94%, but < 100%).
6.   Set the inspiratory time (generally 0.8-1.2 seconds for the adult patient).
7.   Set the level of positive end expiratory pressure (PEEP) to 5 cmH2O (physiologic PEEP). This may be increased if required.
8.   Attach the ventilator circuit to the advanced airway (be sure the capnographic sample line remains in place).
9.   Observe the patient for chest excursion, auscultate breath sounds, and verify the presence of a capnonographic waveform.
10.   Set ventilator alarms:

    • High Pressure Alarm: 10 Points above baseline
    • Low Pressure Alarm: 10 Points below baseline
    • Low Minute Volume Alarm: 10 to 15% below baseline
11.   Observe vital signs, SpO_2/$EtCO_2$, adjust parameters as required. Assess the patient for tolerance and the ventilator system for proper function and adjust accordingly.
12.   Consider the need for sedation and analgesia.

Continuing Mechanical Ventilation from a Sending Facility

1.   Confirm ventilator settings with sending facility staff.
2.   Turn the ventilator on and allow it to self-test/cycle as required for the ventilator being utilized.
3.   Select desired mode of ventilation.
4.   If the patient is receiving volume cycled ventilation, set the desired tidal volume as per sending facility setting.
5.   If the patient is receiving pressure controlled ventilation, set the inspiratory pressure as per the sending facility setting.
6.   Set the inspiratory time as per the sending facility setting (generally 0.8 to 1.0 seconds for the adult patient).
7.   Set the desired rate as per the sending facility setting.
8.   Set the desired FiO2 as per the sending facility sending setting.
9.   Set the level of positive end expiratory pressure (PEEP) as per the sending facility setting.
10.   Set pressure support as per the sending facility setting (if applicable).

P

11. Attach the ventilator circuit to the advanced airway (be sure the capnographic sample line remains in place).
12. Observe the patient for chest excursion, auscultate breath sounds, and verify the presence of a capnographic waveform.
13. If pressure controlled ventilation is being utilized, verify that exhaled tidal volumes are comparable to those achieved during mechanical ventilation by the sending facility.
14. Set ventilator alarms:

   - Volume ventilation: High Pressure Alarm: 10 Points above baseline
     Low Pressure Alarm: 10 Points below baseline
     Low Minute Volume Alarm: 10 to 15% below baseline
   - Pressure ventilation: High Pressure Alarm: 5 Points above baseline
     Low Pressure Alarm: 5 Points below baseline
     Low Minute Volume Alarm: 10 to 15% below baseline

15. Observe vital signs, SpO_2/$EtCO_2$, adjust parameters as required. Assess the patient for tolerance and the ventilator system for proper function and adjust accordingly.
16. Consider need for sedation and ventilation.

Continuing Non-invasive Bi-level Positive Airway Pressure (BIPAP) Support

1. Confirm ventilator settings with sending facility staff.
2. Determine desired exhaled tidal volume (5-8 ml/kg ideal body weight).
3. Determine appropriate mask size.
4. Turn the ventilator on and allow it to self-test/cycle as required for the ventilator being utilized.
5. Set the inspiratory positive airway pressure (IPAP) as per the sending facility setting.
6. Set the expiratory positive airway pressure (EPAP) as per the sending facility setting.
7. Set the FiO_2 as per the sending facility setting.
8. Set the desired backup rate.
9. Connect the patient to the circuit. Adjust the head strap to minimize leaks at the patient-mask interface. Assess the patient for tolerance and the ventilator system for proper function and adjust accordingly.
10. Observe the estimated exhaled tidal volume. This number should approximate the desired exhaled tidal volume as determined above. If the estimated exhaled tidal volume is not maintained close to the desired tidal volume, adjust the patient-unit interface as needed to achieve a steady estimated exhaled tidal volume.
11. Set ventilator alarms as applicable (high pressure, low pressure, time delay, apnea, low minute volume, high rate).
12. Observe vital signs, SpO_2/$EtCO_2$, adjust parameters as required. Assess the patient for tolerance and the ventilator system for proper function and adjust accordingly.

PEARLS:
- High pressure alarm etiologies: kinking/obstruction of the ETT, increased airway resistance (bronchospasm, mucous plugging), decreased lung compliance (pulmonary edema, hyperinflation), decreased thoracic compliance (pneumothorax, gastric distension, abdominal compartment syndrome, patient-ventilator dyssynchrony).
- Low pressure alarm etiologies: circuit disconnect, low ETT cuff pressure, or air leak from drainage and access ports.
- Mandatory modes (AC or SIMV), assure a minimum respiratory rate and tidal volume.

PEARLS:

- Spontaneous modes (pressure-supported ventilation, or PSV) are dependent on patient effort. A minimum number of machine breaths are set by the operator and these are synchronized to the patient's respiratory effort. The patient may take additional breaths, but these will not be supported and could increase work of breathing.
- AC provides full machine support (machine breath) for the programmed rate and for any additional patient-triggered breaths. Hyperventilation is possible.
- Peak pressure is the sum of pressure required to overcome airway resistance and the pressure required to overcome elastic properties of the lung and chest wall. When an inspiratory hold is applied at the end of inspiration, gas flow ceases and the pressure drops to a measurement called the inspiratory plateau or plateau pressure.
- The plateau pressure (Pplat) reflects the pressure required to overcome elastic recoil in the lung and the chest wall, and is the best estimate of peak alveolar pressure reflecting alveolar distention.
- A high Pplat (>30) can be due to excessive tidal volume, gas trapping, PEEP or low compliance. Pplat (alveolar pressure) = (volume/ compliance) + PEEP. The goal is to maintain the Pplat at <30 to decrease this risk of lung injury.
- Optimal measurement of the Pplat requires the absence of patient effort. The Plat is estimated by determining the inspiratory pause pressure, which corresponds to the plateau pressure. The inspiratory pause pressure is determined by activating the inspiratory pause hold function.
- Strategies to reduce the Pplat include decreasing the tidal volume, decreasing extrinsic PEEP, and decreasing auto-PEEP (see below).
- Auto (intrinsic) PEEP results from incomplete expiration prior to the initiation of the next breath. This causes progressive air trapping (hyperinflation) and results in PEEP over and above the level set on the ventilator. This accumulation of air increases alveolar pressure at the end of expiration. Auto-PEEP decreases venous return and can cause hypotension and cardiac arrest (PEA).
- Auto-PEEP can be measured using manual methods or through electronic functions in some ventilators. It is most easily identified by viewing the flow-versus-time graphic available on most ventilators, as the expiratory flow fails to reach the zero flow level before initiation of the next breath.
- Auto-PEEP may be managed by increasing the expiratory time, decreasing the respiratory rate, decreasing the tidal volume, reducing ventilator demand (sedation, treating fever, pain, shivering), or adding external PEEP.
- In bi-level positive airway pressure support, the difference between the EPAP and the IPAP (called the pressure support) augments ventilation and reduces the work of breathing. Increasing the difference between the IPAP and EPAP increases the level of pressure support provided. The difference between the two is adjusted to achieve and effective tidal volume and CO_2 clearance. The EPAP may be adjusted in increments of 2 cmH20 per step to improve oxygenation. A similar increase in the IPAP may be required to maintain the same tidal volume.
- For patients with obstructive pulmonary disease, evaluate the patient for auto-PEEP and manage inspiratory time and expiratory time to reduce or eliminate auto-PEEP. Patients with obstructive pulmonary disease will require a longer expiratory time.
- Asthmatics can be particularly sensitive to the hemodynamic effects of positive pressure ventilation and are at increased risk for developing auto-PEEP. Minimizing hyperinflation and avoiding excessive airway pressures are key goals in ventilating these patients. These goals are best accomplished by utilizing low respiratory rate and tidal volume to allow sufficient time for exhalation. Higher than normal CO_2 levels should be tolerated in these patients.

HEIGHT	PBW	4 ml	5 ml	6 ml	7 ml	8 ml
4' 0" (48)	17.9	72	90	107	125	143
4' 1" (49)	20.2	81	101	121	141	162
4' 2" (50)	22.5	90	113	135	158	180
4' 3" (51)	24.8	99	124	149	174	198
4' 4" (52)	27.1	108	136	163	190	217
4' 5" (53)	29.4	118	147	176	206	235
4' 6" (54)	31.7	127	159	190	222	254
4' 7" (55)	34	136	170	204	238	272
4' 8" (56)	36.3	145	182	218	254	290
4' 9" (57)	38.6	154	193	232	270	309
4' 10" (58)	40.9	164	205	245	286	327
4' 11" (59)	43.2	173	216	259	302	346
5' 0" (60)	45.5	182	228	273	319	364
5' 1" (61)	47.8	191	239	287	335	382
5' 2" (62)	50.1	200	251	301	351	401
5' 3" (63)	52.4	210	262	314	367	419
5' 4" (64)	54.7	219	274	328	383	438
5' 5" (65)	57	228	285	342	399	456
5' 6" (66)	59.3	237	297	356	415	474
5' 7" (67)	61.6	246	308	370	431	493
5' 8" (68)	63.9	256	320	383	447	511
5' 9" (69)	66.2	265	331	397	463	530
5' 10" (70)	68.5	274	343	411	480	548
5' 11" (71)	70.8	283	354	425	496	566
6' 0" (72)	73.1	292	366	439	512	585
6' 1" (73)	75.4	302	377	452	528	603
6' 2" (74)	77.7	311	389	466	544	622
6' 3" (75)	80	320	400	480	560	640
6' 4" (76)	82.3	329	412	494	576	658
6' 5" (77)	84.6	338	423	508	592	677
6' 6" (78)	86.9	348	435	521	608	695
6' 7" (79)	89.2	357	446	535	624	714
6' 8" (80)	91.5	366	458	549	641	732
6' 9" (81)	93.8	375	469	563	657	750
6' 10" (82)	96.1	384	481	577	673	769
6' 11" (83)	98.4	394	492	590	689	787
7' 0" (84)	100.7	403	504	604	705	806

PBW and Tidal Volume for Females

HEIGHT	PBW	4 ml	5 ml	6 ml	7 ml	8 ml
4' 0" (48)	22.4	90	112	134	157	179
4' 1" (49)	24.7	99	124	148	173	198
4' 2" (50)	27	108	135	162	189	216
4' 3" (51)	29.3	117	147	176	205	234
4' 4" (52)	31.6	126	158	190	221	253
4' 5" (53)	33.9	136	170	203	237	271
4' 6" (54)	36.2	145	181	217	253	290
4' 7" (55)	38.5	154	193	231	270	308
4' 8" (56)	40.8	163	204	245	286	326
4' 9" (57)	43.1	172	216	259	302	345
4' 10" (58)	45.4	182	227	272	318	363
4' 11" (59)	47.7	191	239	286	334	382
5' 0" (60)	50	200	250	300	350	400
5' 1" (61)	52.3	209	262	314	366	418
5' 2" (62)	54.6	218	273	328	382	437
5' 3" (63)	56.9	228	285	341	398	455
5' 4" (64)	59.2	237	296	355	414	474
5' 5" (65)	61.5	246	308	369	431	492
5' 6" (66)	63.8	255	319	383	447	510
5' 7" (67)	66.1	264	331	397	463	529
5' 8" (68)	68.4	274	342	410	479	547
5' 9" (69)	70.7	283	354	424	495	566
5' 10" (70)	73	292	365	438	511	584
5' 11" (71)	75.3	301	377	452	527	602
6' 0" (72)	77.6	310	388	466	543	621
6' 1" (73)	79.9	320	400	479	559	639
6' 2" (74)	82.2	329	411	493	575	658
6' 3" (75)	84.5	338	423	507	592	676
6' 4" (76)	86.8	347	434	521	608	694
6' 5" (77)	89.1	356	446	535	624	713
6' 6" (78)	91.4	366	457	548	640	731
6' 7" (79)	93.7	375	469	562	656	750
6' 8" (80)	96	384	480	576	672	768
6' 9" (81)	98.3	393	492	590	688	786
6' 10" (82)	100.6	402	503	604	704	805
6' 11" (83)	102.9	412	515	617	720	823
7' 0" (84)	105.2	421	526	631	736	842

PBW and Tidal Volume for Males

A
C
P

Indication:
- Patient with a complaint of chest pain/discomfort consistent with a cardiac etiology or other known or suspected anginal equivalent.
- Any patient 35 years old with a complaint of chest pain (non traumatic), chest pressure, chest tightness, non-specific chest discomfort, heartburn, syncope, difficulty breathing (without obvious respiratory etiology), palpitations, or signs or symptoms of congestive heart failure.
- Adult female with complaint of weakness, dizziness or nausea.
- Any patient of any age with any of the above symptoms AND a history of diabetes, cardiac disease, obesity, recent cocaine use, or a family history of premature cardiac disease.

Contraindications:
- None.

Background:
Prehospital acquisition of 12-lead electrocardiograms (ECGs) has been recommended by the AHA Guidelines for CPR and Emergency Cardiovascular Care since 2000. Obtaining an ECG during the prehospital phase of care in the patient with possible acute coronary syndrome allows for the early identification of ST elevation myocardial infarction (STEMI), triage of the STEMI patient to a primary PCI (PPCI) capable facility, and early notification to the receiving facility (activation of the cardiac catheterization laboratory [CCL]) resulting in a reduction in both the time to reperfusion and mortality. The multilead ECG is also potentially helpful in identifying dysrhythmias, conduction disturbances, metabolic and electrolyte disorders, cardiac structural abnormalities, pulmonary embolism and changes associated with toxic ingestions.

Procedure:
1. Prepare ECG monitor and connect patient cable with electrodes.
2. Enter required patient information (patient ID, DOB etc.) into the ECG device.
3. Expose the chest and prep as necessary (modesty of the patient should be respected).
4. Apply extremity electrodes/leads using the following landmarks:
   ○ RA - Right arm
   ○ LA - Left arm
   ○ LL - Left leg
   ○ RL - Right leg
   ○ V1 - 4th intercostal space at the right sternal border
   ○ V2 - 4th intercostal space at the left sternal border
   ○ V3 - Directly between V2 and V4
   ○ V4 - 5th intercostal space at the midclavicular line
   ○ V5 - Level with V4 at the anterior axillary line
   ○ V6 - Level with V5 at the left mid-axillary line
5. Instruct the patient to remain still.
6. Press the device specific button to acquire a 12 lead ECG.
7. Depending on provider level, interpret the ECG or transmit the ECG for interpretation (if the ECG is transmitted, notify the receiving facility of the transmission as soon as possible).
8. Declare CODE STEMI to receiving facility if indicated.
9. Manage patient per appropriate protocol (s).
10. Print two copies of the ECG. One copy is to be left with the receiving facility and the other is to be retained by the EMS service as part of the patient's medical record. If the ECG interpreted by the EMS provider, the interpretation must be documented on the patient care report.

Conventional Lead Placement for a 12 Lead ECG:

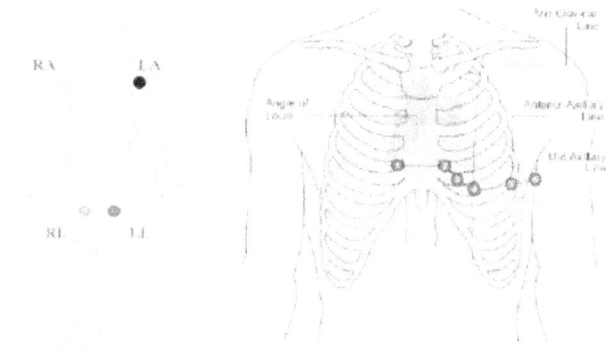

Lead Placement for Posterior ECG:

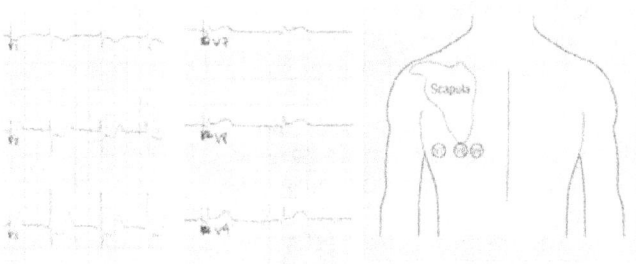

A C P

PEARLS:

- Serial (repeat ECGs) should be performed with the adhesive electrodes in the same location as the previous ECG(s) and the ECG should be acquired with the patient in the same position.
- Be alert for causes of artifact: dry skin, diaphoresis, dried out electrodes, patient or cable movement, electromagnetic interference.
- Electrodes should be stored in the original foil sealed packaging to prevent drying.
- Do not mix brands or types of electrodes when acquiring an ECG.
- Diaphoretic patients should be dried as much a possible. Consider the use of tincture of benzoin to aid in electrode adherence.
- Right ventricle myocardial infarction is present in up to 50% of inferior wall infarctions. A right sided ECG should be considered in patients with inferior wall STEMI. A right-sided ECG is a "mirror reflection" of the standard left sided 12-lead ECG. Connect the V1 lead cable to electrode V1R, continue connecting lead cables to electrodes in sequence until lead cable V6 is connected to electrode V6R. STE > 1mm in lead V4R is sensitive for RV infarction (88-100% sensitivity, 78-82% specificity, 83-92% diagnostic accuracy).
- Consider a posterior ECG (to identify posterior wall infraction) in patients with Inferior or lateral wall MI (especially if accompanied by ST depression or prominent R waves in leads V1-V3). To obtain, place electrodes as follows: V9 - left spinal border, same horizontal line as V4-6; V8 - midscapular line, same horizontal line as V7 and V9, and V7; posterior axillary line, same horizontal line as V4-6. Connect leads to the electrodes as follows: V6 to V9, V5 to V8, and V4 to V7. ST elevation > 0.5mm in leads V8-9 is sensitive for posterior wall infarction (as high as 90, with predictive accuracy up to 93.8%).

Indication:
- Supraventricular tachycardia (HR>150, regular rhythm).

Contraindications:
- Patient age > 65.
- History of transient ischemic attack or cerebrovascular accident (in the last three months) or carotid artery vascular disease/surgery.
- Presence of a carotid bruit.

Background:
Carotid Sinus Massage (CSM) triggers the baro receptor reflex and results in increased vagal tone affecting the sinoatrial and atrioventricular nodes. CSM sometimes terminates supraventricular tachycardia.

Procedure:
1. Assess patient for contraindications as above.
2. Initiate ECG monitoring.
3. Explain procedure to the patient.
4. Identify the carotid sinus location at midpoint between the angle of the mandible and the superior border of thyroid cartilage.
5. Massage the right carotid sinus 5 seconds while observing the ECG (massage firmly but gently, use the same pressure that would indent tennis ball, do not apply so much pressure to occlude carotid).
6. If there no response from massaging the right carotid sinus, massage the left carotid sinus for 5 seconds while observing the ECG.
7. Massage should be stopped immediately if neurological symptoms of any kind, syncope AV block or asystole develop.

P
8. Document procedure and patient response.

Indication:
- Patient in cardiac arrest.

Contraindications:
- None.

Background:
Defibrillation is the unsynchronized random delivery of electricity to the myocardium. Defibrillation results in widespread depolarization of myocardial cells. This widespread depolarization can terminate ventricular fibrillation or ventricular tachycardia allowing the myocardium to regain normal electrical activity. When indicated, defibrillation should be performed immediately and without delay. In witnessed ventricular fibrillation, for every minute that passes between the onset of ventricular fibrillation and defibrillation, the survival rate decrease 7% to 10%. When bystander CPR is provided, the decrease in the survival rate is more gradual and averages 3% to 4 % per minute. The delivery of unsynchronized electricity (defibrillation) is sometimes necessary when the delivery synchronized electricity (synchronized cardioversion) is not possible due to failure to sense (R waves) in the unstable patient with other tachydysrythmias. Automated external defibrillators (AEDs) allow non-ALS EMS providers and members of the lay public to provide timely defibrillation.

E

Procedure:
1. Ensure that chest compressions are adequate and interrupted only when absolutely necessary. If multiple providers are available, one provider should provide uninterrupted chest compressions while the AED is being prepared for use.
2. Apply electrical therapy pads per manufacturer recommendations. Use alternate placement configuration when implanted devices (implanted defibrillator or pacemaker) occupy preferred pad positions. If the patients age is <8 yo, use pediatric pads if available or if device has "energy attenuation" function key, activate the key.
3. If present, remove any transdermal medication patches and wipe off any residue.
4. Power on the AED and follow prompts.
5. Keep interruption of chest compressions to a minimum.
6. If shock is advised, assertively state "All Clear" and visualize that everyone (including yourself) is clear and press the shock button.
7. If no shock is advised, immediately resume chest compressions.
8. Allow the AED to analyze when prompted (approximately two minute cycles). Repeat steps 6 and 7 as indicated.

PEARLS:
- Pre and post-shock pauses should be minimized! Shorter perishock pauses (specifically preshock) are associated with greater survival from out-of-hospital cardiac arrest.
- If pads touch due to patient size or use of a Lucas Device, use the AP configuration for pad placement.
- For patients with implanted devices (defibrillators/pacemakers), pads can be placed in the AL or AP configuration. Attempt to avoid placing pads directly over devices.
- In the AP configuration, one pad is placed over the left anterior mid chest next to the sternum and another pad is placed to the mid left posterior chest next to the spine.
- For patients with refractory ventricular fibrillation, consider replacing or changing therapy pads with new ones and ensure good skin contact. Consider AP placement if AL placement is being used.

Indication:
- Ventricular fibrillation or pulseless ventricular tachycardia.

Contraindications:
- None

Background:
Defibrillation is the unsynchronized random delivery of electricity to the myocardium. Defibrillation results in widespread depolarization of myocardial cells. This widespread depolarization can terminate ventricular fibrillation or ventricular tachycardia allowing the myocardium to regain normal electrical activity. When indicated, defibrillation should be performed immediately and without delay. In witnessed ventricular fibrillation, for every minute that passes between the onset of ventricular fibrillation and defibrillation, the survival rate decrease 7% to 10%. When bystander CPR is provided, the decrease in the survival rate is more gradual and averages 3% to 4 % per minute. The delivery of unsynchronized electricity (defibrillation) is sometimes necessary when the delivery synchronized electricity (synchronized cardioversion) is not possible due to failure to sense (R waves) in the unstable patient with other tachydysrrythmias.

Procedure:
1. Ensure that chest compressions are adequate and interrupted only when absolutely necessary.
2. If not in place, apply electrical therapy pads in the anterior-lateral (AL) configuration. Place the lateral (apex) pad on the left anterior chest, just lateral to the left nipple in the mid-axillary line and the anterior pad in the right sub-clavicular area lateral to the sternum (the anterior-posterior [AP] configuration may be used if pads are already in place in this configuration). If present, chest hair should be shaved prior to pad application.
3. Select appropriate energy level.
4. Charge the defibrillator to selected energy level. Chest compressions should be continued while the defibrillator is being charged.
5. Once the defibrillator is charged, hold chest compressions, assertively state "All Clear" and visualize that everyone (including yourself) is clear.
6. Push the discharge button to deliver the countershock.
7. Once the countershock is delivered, immediately resume chest compressions (it may be desirable to have the chest compressor "hover" their interlocked hands over the chest while the countershock is delivered to minimize the length of postshock pause).
8. Continue chest compressions and ventilations for two minutes. After two minutes of CPR, analyze the rhythm and check for pulse only if appropriate for the rhythm.
9. Repeat defibrillation every two minutes as indicated by the ECG and patient response.
10. Document procedure, including ECG rhythm, energy level and response.

A C P

PEARLS:

- Pre and post-shock pauses should be minimized! Shorter perishock pauses (specifically preshock) are associated with greater survival from out-of-hospital cardiac arrest. In one study, preshock pauses >20 sec had a 53% lower chance of survival compared to those with preshock pauses <10 sec. For every 5 sec increase in shock pause, the chance of survival decreased by 18%.
- If pads touch due to patient size or use of a Lucas Device, use the AP configuration for pad placement.
- For patients with implanted devices (defibrillators/pacemakers), pads can be placed in the AL or AP configuration. Attempt to avoid placing pads directly over devices.
- In the AP configuration, one pad is placed over the left anterior mid chest next to the sternum and another pad is placed to the mid left posterior chest next to the spine.
- For patients with refractory ventricular fibrillation, consider replacing changing therapy pads with new ones and ensure good skin contact. Consider AP placement if AL placement is being used.

A

C

P

Indication:
- Adult patient with ventricular fibrillation (VF) or pulseless ventricular tachycardia (PVT) refractory to ≥3 defibrillation attempts (AED or manual) **and** one dose of epinephrine **and** one dose of a first line antiarrhythmic agent (Amiodarone or lidocaine).

Contraindications:
- None.

Background:
Double sequential defibrillation (DSED) is the simultaneous delivery of two high energy defibrillation doses. DSED requires the presence of two manual biphasic defibrillators. DSED was first described in the 1980s. The exact mechanism by which DSED works is unclear. It is believed that the delivery of two simultaneous shocks via two different vectors decreases the total energy and peak voltage required to terminate ventricular fibrillation. For the purposes of this protocol, refractory VF/PVT is defined as VF/PVT refractory to ≥3 defibrillation attempts and one dose of epinephrine and one dose of a first line antiarrhythmic agent (Amiodarone or Lidocaine). Refractory VF/PVT must be differentiated from recurrent VF/PVT. Recurrent VF/PVT is VF/PVT that reoccurs episodically after successful conversion with intervening episodes of organized electrical activity. DSED is not indicated for recurrent VF/PVT. It should be noted that a patient with recurrent VF/PVT may develop refractory VF/PVT. While the current literature supporting the use of DSED is limited to case reports, there is little to no reason to not consider its use in the patient with refractory VF/PVT.

Procedure:
1. Apply a second set of electrical therapy pads in an alternate configuration (two sets of pads are applied, one set in the anterior-apical configuration and the other in the anterior-posterior configuration). The site of application should be dried and hair   should be removed to maximize pad adherence.
2. Select the maximum energy level on both defibrillators.
3. Charge both defibrillators simultaneously. Both should be fully charged 15 seconds prior to the anticipated pause in chest compressions. Assure the continuance of high quality chest compressions during charging.
4. At the prescribed time in the compression cycle, pause compressions and discharge the defibrillators simultaneously. The time from pause to shock delivery should be minimized.
6. Immediately resume high quality chest compressions. After a 2 minute compression cycle, reevaluate the ECG rhythm.
7. Repeat the procedure as indicated. All subsequent defibrillations should be double sequential.

PEARLS:
- DSED requires the presence of two defibrillators. Providers should anticipate the need for a second defibrillator early in the course of resuscitation if DSED is indicated.
- For antero-apical placement, one pad is placed to the right of the sternum just below the clavicle, and the other is centered lateral to the normal cardiac apex in the anterior or midaxillary line (V5–6).
- For antero-posterior placement, the anterior pad is placed over the precordium or apex, and the posterior pad is placed on the posterior thorax in the left infrascapular region.
- DSED should be documented at such on the PCR.

A
C
P

Indication:
- Unstable patient with a tachydysrhythmia (atrial fibrillation with rapid ventricular response, supraventricular tachycardia, ventricular tachycardia).
- Patient is not pulseless (the pulseless patient requires unsynchronized cardioversion i.e. defibrillation).

Contraindications:
- Repetitive, self-terminating, short-lived tachydysrhythmias (i.e. runs of non-sustained VT).

Background:
In synchronized cardioversion the delivery of electricity is synchronized with the cardiac cycle so that delivery occurs during the absolute refractory period. This synchronization avoids shock delivery during the relative refractory portion of the cardiac electrical cycle, when a shock could produce ventricular fibrillation. Cardioversion is useful in the treatment of tachydysrhythmia (atrial fibrillation with rapid ventricular response, supraventricular tachycardia, ventricular tachycardia). Energy doses utilized for cardioversion are generally lower than those used for defibrillation.

Procedure:
1. Explain procedure to patient if applicable.
2. Ensure standard resuscitation equipment is immediately available.
3. Consider pre-procedure sedation per the *Patient Comfort Protocol* if appropriate for the clinical situation and provider level scope of practice.
4. Attach ECG monitoring electrodes in the standard 4 lead configuration.
5. The preferred configuration for application of electrical therapy pads is the anterior-posterior (AP) configuration. In the AP configuration, one pad is placed to the anterior mid chest next to the sternum and another pad is placed to the mid left posterior chest next to the spine. Alternately, the anterior-lateral (AL) configuration may be utilized. In the anterior-lateral configuration, the lateral (apex) pad is placed on the left anterior chest, just lateral to the left nipple in the mid-axillary line and the anterior electrode is placed in the right subclavicular area lateral to the sternum.
6. Set the defibrillator to synchronized cardioversion mode and observe for R wave markers. If the R wave markers do not appear, or appear elsewhere on the ECG, select another lead or reposition the ECG electrodes. Increasing the gain to increase QRS amplitude may also be helpful.
7. Select desired energy level, charge the device and assertively state "All Clear" and visualize that everyone (including yourself) is clear.
8. Depress and hold the energy discharge button until discharge occurs (there may be a delay between depressing the energy discharge button and discharge).
9. Reassess the patient and the ECG rhythm. If the rhythm is unchanged and the patient's condition is the same, repeat with escalating energy levels per appropriate protocol. In the event the patient's rhythm deteriorates into ventricular fibrillation or pulseless ventricular tachycardia, immediately perform unsynchronized cardioversion (defibrillation).
10. Document procedure, including ECG rhythm, energy level and response.

PEARLS:

- In the event cardioversion is unsuccessful after multiple attempts, consider replacing or changing therapy pads with new ones and ensure good skin contact. Consider AP placement if AL placement is being used.
- For patients with implanted devices (defibrillators/pacemakers), pads can be placed in the AL or AP configuration. Attempt to avoid placing pads directly over devices.
- In the AP configuration, one pad is placed over the left anterior mid chest next to the sternum and another pad is placed to the mid left posterior chest next to the spine.
- Delivering electrical therapy at end expiration may decrease transthoracic resistance.

A
C
P

Indication:
- Bradycardia with inadequate cardiac or cerebral perfusion as evidenced by hypotension, altered mental status, chest pain/discomfort, pulmonary edema, or other signs of shock.

Contraindications:
- Hypothermia with a core temperature < 86° F.

Background:
Heart rate is a component of cardiac output (CO) [HR X SV = CO]. Many patients tolerate bradycardia (HR < 60 in the adult) and experience no hemodynamic compromise, but some patients with profound bradycardia or those that are dependent on heart rate to maintain cardiac output may experience decreased cardiac or cerebral perfusion as a result of bradycardia. TCP is an option for treating bradycardia when it associated with hemodynamic compromise (decreased perfusion). TCP is considered equivalent to pharmacotherapy (atropine sulfate) and may be used immediately (in lieu of atropine sulfate) or in patients who do not respond to atropine sulfate. It should be noted that when there is impairment in the conduction system resulting in a high-degree block (e.g., Mobitz type II second-degree block or third-degree AV block), atropine sulfate is unlikely to be effective.

Procedure:
1. Explain procedure to patient if applicable.
2. Attach ECG monitoring electrodes in the standard 4 lead configuration.
3. Apply pacing/multifunction therapy pads. Anterior-Posterior (AP) application is preferred for pacing. In the AP configuration, one pad is placed to the anterior mid chest next to the sternum and another pad is placed to the mid left posterior chest next to the spine. Alternately, the Anterior-Lateral (AL) configuration may be utilized. In the anterior-lateral configuration, the lateral (apex) pad is placed on the left anterior chest, just lateral to the left nipple in the midaxillary line and the anterior electrode is placed in the right subclavicular area lateral to the sternum.
4. Select pacing function on the monitor unit.
5. Set pacing rate to 70 bpm for an adult patient and 100 bpm for a pediatric patient.
6. Note pacing spikes on the screen. Beginning at the lowest mA setting possible, gradually increase the current until electrical capture occurs. In most patients capture will be achieved with an output of 50-100 mA.
7. Once capture electrical capture is achieved, check for corresponding mechanical capture as evidenced by a palpable pulse or presence of an oximetric (arterial) waveform.
8. Choose demand mode and verify sensing. The non-demand mode may be used when inhibition due to sensing of signals other than R waves, such as muscle artifact, or P or T waves (over sensing) and other troubleshooting measures are unsuccessful.
9. Assess patient response, including vital signs in response to pacing.
10. Consider the use of sedation and or analgesia if indicated per age appropriate *Patient Comfort Protocol*.
11. Document procedure including indication (initial ECG rhythm and hemodynamic status), pacer settings (mode, mA, rate) and the patient's response to pacing.

PEARLS:

- Adult pacing/multifunction therapy pads may be used in patients > 15 kg (33 lbs) or 1 yo.
- Avoid placement of pads over bony prominences of the spine or scapula.
- Ensure good skin contact with pad application. Shave hair if situation permits.
- Electrical capture is usually represented by a widening of the QRS complex and a tall, broad T wave. The deflection of the captured complex may be positive or negative.
- If electrical capture occurs without corresponding mechanical capture, consider/rule out the following: hypovolemia, tension pneumothorax, pericardial tamponade, pulmonary embolus, extensive myocardial infarction, or profound acidosis.
- Strong muscular contractions may make it difficult to accurately palpate a pulse. When pacing with the electrodes in the AP position, palpation of the carotid, brachial or femoral artery should be done on the patient's right side and blood pressure should also be measured on this side. Utilize an oximetric (arterial) waveform to assist in identifying mechanical capture.

Indication:
- Cardiac arrest.

Contraindications:
- LUCAS© patient <12 yo, AutoPulse™ and ROSC-U patient <18 yo.
- Obviously pregnant patient.
- Traumatic cardiac arrest.
- Device does not fit patient.

Background:
Currently available mechanical CPR devices include piston driven devices or load distributing band devices (LDB). To date there is no literature demonstrating outcomes related superiority of these devices when compared to manually performed CPR. The 2015 AHA Guidelines Update for CPR and ECC state "The use of LDB-CPR may be considered in specific settings where the delivery of high-quality manual compressions may be challenging or dangerous for the provider (e.g., limited rescuers available, in a moving ambulance), provided that <u>rescuers strictly limit interruptions in CPR during deployment and removal of the devices</u>. The use of these devices are classified as a Class IIb recommendation (usefulness/efficacy is less well established by evidence/opinion). The procedure for the use of two common devices is detailed below. EMS providers may utilize other FDA approved mechanical CPR devices. The validation and documentation of the individual EMS provider's competency utilizing a particular device is ultimately the responsibility of the service Medical Director.

Procedure:

<u>LUCAS©</u>

E

1. Ensure the operation knob is in the ADJUST position.
2. Assemble/prepare the device as per type being used (electric or pneumatic).
3. At the time of the next 2 minute CPR duty cycle pause, apply a posterior electrical therapy/defibrillation pad and place the patient on a backboard.
4. Place the back plate under the patient on backboard below the armpits.
5. Resume chest compressions.
6. Position the suction cup so the lower edge is immediately above the end of the sternum and the pressure pad is centered over the middle of the sternum.
7. Lower the suction cup and pressure pad to the point where it is just comes into contact with the patient's lower chest. If the pad does not fit, resume manual chest compressions.
8. Turn the operation knob to ACTIVE.
9. Check the device for proper position.
10. Attach the stabilization straps.
11. To stop LUCAS operation, turn the operation knob to LOCK (this should only be done if the device is improperly placed, injury to the patient is occurring, to assess the patient, or when an AED is analyzing and charging).
12. If sustained return of spontaneous circulation is achieved, release and retract the "pressure pad" to allow for greater chest excursion and tidal volume during bag-valve-mask ventilation.

AutoPulse™

1. Power on the AutoPulse platform and allow the device to initialize and perform the device self-test. Proceed only if the device indicates it is ready for use.
2. At the time of the next 2 minute CPR duty cycle pause, sit the patient up and cut any clothing by making a single cut down the back.
3. Slide the AutoPulse platform into position behind the sitting patient and lay the patient down onto the platform.
4. Grasp any clothing that was cut by the sleeves and pull downward toward the ankles to remove all clothing from both the front and back of the torso.
5. CPR may be resumed for a two minute duty cycle depending on the time expended thus far. If this is the case, resume deployment at the end of the duty cycle.
6. Position the patient so that the patient is centered laterally (from left to right) and that the patient's armpits are aligned with the AuoPulse platform using the yellow line positioning guides on the AutoPulse platform.
7. Align the chest bands. Place band 1 on top of the patient's chest, locate the mating slot of band 2 over the alignment tab. Press the bands together to engage and secure the Velcro fastener (make sure the bands are not twisted before automatic compressions begin, if the bands cannot be closed, use manual CPR).
8. Make sure that the yellow upper edge of the chest bands are aligned with the patient's armpits, and that it is directly over the yellow line on the platform. There should be no obstructions (clothing, straps, or equipment) with the bands.
9. Press and release the START-CONTINUE button once (the AutoPulse will automatically adjust the bands to the patient's chest).
10. Verify the patient is properly aligned and that all slack in the bands has been taken up.
11. Press and release the START-CONTINUE button a second time. The AutoPulse will begin automated compressions.
12. The user has the option of switching between compressions delivered at 15:2 or continuous.
13. Positive pressure ventilation can be performed synchronous with any decompression and or with the ventilation pause. Check the patient's chest rise during active operation.
14. Compressions may be paused by pressing the STOP-CANCEL button, push RESUME to resume compressions.
15. To turn the AutoPulse off, press the STOP-CANCEL button followed by the ON-OFF button.

ROSC-U MCCR©

1. Ensure high quality manual chest compressions are ongoing.
2. Prepare the unit for application.
3. Ensure the Main Power Switch on Battery Control Unit is in the ON position which is indicated by it being illuminated.
4. Prepare to apply the device at the time of the next anticipated pause.
5. Roll the patient on to his/her side and slip the torso restraint under patient's back.
6. Position the center of the compressor along an imaginary line connecting the nipples approximately 3-4 cm/1.2-1.6 in (1-2 fingers) from bottom of sternum notch.
7. Wrap the torso restraint around patient and secure the compressor to the patient's body by firmly pulling fabric ends of the torso restraint fully out (creating tension) then down onto patient's chest.
8. Verify that the device is properly positioned and secured on the patient.
9. Turn the power ON using the Main Green Power Rocker Switch (it will illuminate).
10. Start compressions by pushing the Red Power Button.
11. If desired, adjust piston depth beyond default depth using the Yellow Piston Depth Set Button.
12. Pause/restart compressions by pushing the Green Run/Pause Button.
13. When resuscitation efforts are terminated, push the Red Power ON/OFF button to stop compression cycles and then turn Main ON/OFF Power Rocker Switch to the OFF position. Disconnect the Torso Restraint from the compressor. Remove the compressor from patient's chest.

E

PEARLS:
- Deployment or removal of external CPR devices should not significantly (> 10 seconds) interrupt chest compressions.
- System personnel should train with their respective devices to minimize deployment time and interruptions in chest compressions.
- External CPR devices should never be left unattended or with an untrained provider.
- Providers should be familiar with the specific device being utilized and should receive formal in-service training provided by their service specific to the device.

Indication:
- Patient with altered mental status, signs/symptoms of hypoglycemia or diabetic ketoacidosis, patients with metabolic or endocrine disorders presenting with non-specific complaints, stroke assessment, and infants with hypothermia or bradycardia.

Contraindications:
- None

Background:
Hypoglycemia is a potentially life threatening condition. The use of a glucometer for blood glucose analyses allows for the rapid determination of the blood glucose (bG) level in the field.

Procedure:
1.   Explain procedure to patient if applicable.
2.   Blood samples for performing glucose analysis should be obtained using a capillary blood sample from a finger-stick or the heel in an infant.
3.   Prepare site with alcohol or chlorohexidine.
4.   Utilize lancing device and then place the correct amount of blood on the reagent test strip or on the glucometer per the manufactures recommendation.
5.   Time the analysis per the manufactures recommendation.
6.   Document bG reading and treat patient as indicated per age appropriate _Diabetic Emergencies Protocol._

E

PEARLS:
- QA should be performed on glucometers at least once every 7 days or as recommended by the manufacture and in the event clinically suspicious readings are noted. Documentation of QA must be maintained.

Indication:
- Patient requiring continuous core temperature monitoring.

Contraindications:
- Patient not requiring continuous core temperature monitoring.

Background:
Placement of an esophageal temperature probe will allow monitoring of an approximated core body temperature. The distal tip of a properly placed esophageal temperature probe is located in the proximity of the aorta and heart. If the probe is placed too high in the esophagus, the reading will be affected by tracheal air.

Procedure:
1. Connect the esophageal temperature probe to a temperature monitor.
2. Observe the measured temperature to determine if it closely matches the ambient room temperature.
3. Determine the appropriate depth of insertion by measuring from the nares or the corner of the mouth, depending on which site is to be used, to 2 cm above the xiphoid process.
4. Lubricate the distal portion of the probe with water soluble lubricant.
5. Insert the tube via the nose or mouth to the predetermined depth.
6. Secure the tube in place with tape.

P

Indication:
- Suspected tension pneumothorax.
- Patient in traumatic cardiac arrest with chest or abdominal trauma with or without signs suggestive of tension pneumothorax.

Contraindications:
- None

Background:
Tension pneumothorax may result from blunt or penetrating trauma. If allowed to progress, tension pneumothorax results in cardiovascular collapse. It is clinically identified by the presence of respiratory distress, hyperresonance to percussion on the affected side, decreased or absent breath sounds, jugular venous distention and hypotension. Tracheal deviation away from the affected side may be seen or palpated, but is a late clinical finding. In the intubated patient, decreased lung compliance may be noted. Tension pneumothorax should be considered in all cases of pulseless electrical activity (PEA) arrest.

Procedure:

P

1. Position the patient in a supine position (if possible given situation and patient location).
2. Identify insertion site: 2nd intercostal space in the midclavicular line or the 4th-5th intercostal space in the anterior axillary line (the anterior axillary line location is preferred in obese patients).
3. Prepare the site with alcohol or chlorhexidine.
4. Insert a 14g 3.25" (or longer) decompression device or catheter over the needle device into the skin over the top of the rib located just below the intercostal space of choice.
5. Advance the device through the parietal pleura (a distinctive "pop" should be noted) until air or blood exits (do not continue to advance the needle at this point as there is risk of injury to the lung during re-expansion).
6. Advance the catheter (not the needle) to the chest wall.
7. Remove the needle, leaving the plastic catheter in place.
8. If available, a one way valve may be attached.
9. Document the procedure including clinical signs suggestive of tension pneumothorax, location of procedure, type and size of device used, and clinical response.

PEARLS:
- In the neonate, utilize a 23 or 25 g butterfly needle or a 20 or 22g catheter over the needle device. Both can be attached to a three-way stopcock and a syringe (a T-connector will be needed to attach a catheter over the needle device).
- In the pediatric patient, utilize a 16 or 18 g catheter over the needle device.
- If the patient does not improve, consider decompressing the opposite side.
- Some patients may require multiple decompressions on the same side.

Indication:
- Ocular irrigation after chemical exposure/thermal injury.
- Facilitate removal of non-embedded foreign material from the eye.

Contraindications:
- Patient < 8 years of age.

Background:
The Morgan Lens© is a sterile plastic device resembling a contact lens that fits over the eye similar to a contact lens. The device connects to irrigation tubing. The device allows for copious irrigation of the eye(s).

Procedure:
1. Instill topical ophthalmic anesthetic in to the affected eye(s).
2. Mix 100 mg of LIDOCAINE (5ml of a 2% solution) in 1000 ml LACTATED RINGER'S SOLUTION.
3. Attach the Morgan Lens© to a delivery set equipped with a macro drip chamber and open the flow control to start flow.
4. Instruct the patient to look down and insert the upper portion of the lens under the upper eye lid.
5. Instruct the patient to look up and retract the lower lid allowing placement of the lower portion of the lens under the lower lid.
6. Continue irrigation of the affected eye(s) using caution to ensure run off does not enter the unaffected eye. Do not allow the irrigation solution to run dry.
7. Tape the tubing to the patient's face to prevent inadvertent removal.
8. Consider additional pain management as indicated.
9. To remove the lens, continue the flow of irrigation solution while instructing the patient to look up. Retract the lower lid and slide the lens from the upper lid.
10. All patients should be transported for evaluation for corneal injury.

P

PEARLS:
- Once topical ophthalmic anesthetic is instilled, care should be taken that the patient doesn't rub his/her eyes as damage can occur.

Indication:
- Gastric decompression in the intubated patient.
- Patient at high risk for aspiration due to vomiting.
- Burns >20 body surface area (BSA).

Contraindications:
- History of gastric bypass surgery or recent gastric banding.
- Caustic substance ingestion.
- Maxillofacial trauma (nasal route).
- Patient on warfarin or other anticoagulant/antiplatelet agents (nasal route, relative contraindication).

Background:
Gastric decompression is indicated in all intubated patients. Additionally, gastric intubation/decompression should be considered in patient at high risk for aspiration due to vomiting and those with burns affecting >20 bsa. When possible (intubated patient usually), the oral route is preferred over the nasal route due to the risk for infection. In the intubated patient, gastric decompression is a part of airway management as it decreases the risk for micro aspiration. In cardiac arrest gastric decompression eliminates the potential deleterious effects associated with gastric distention.

Procedure:

Oral Route
1. Select appropriate tube size (16-18fr for most adults, pediatric tube size may be estimated by the following formula: Age in years + 18 /2 (8 yo patient example: 8+16 = 24/2 = 12fr).
2. Estimate the insertion length by superimposing the tube over the patient's body from the xiphoid to the angle of the jaw to the corner of the mouth (the insertion depth can be marked with tape).
3. Depress the tongue and insert the tube into the oropharynx.
4. Continue to advance the tube (the tube should be passed with a minimum of pressure, with laryngoscope assistance if necessary).
5. Advance the tube until gastric contents are returned or the desired depth of insertion is reached (if coughing occurs, halt insertion, retract the tube and reattempt insertion).
6. Confirm placement by auscultating for gastric sounds over the epigastrium while insufflating 20ml of air. Additionally, confirm placement by aspirating gastric contents.
7. Decompress the stomach of air and gastric contents by attaching the gastric tube to low suction or by manually aspirating with a large (Toomey) catheter tip syringe.
8. Secure the tube in place with tape.
9. Document gastric tube placement and confirmation by auscultation/aspiration of gastric content. The volume of liquid output should also be documented.

P

Nasal Route

1. Select appropriate tube size (16-18fr for most adults, pediatric tube size may be estimated by the following formula: Age in years + 18 /2 (8 yo patient example: 8+16 = 24/2 = 12fr).
2. Estimate the insertion length by superimposing the tube over the patient's body from the xiphoid to the angle of the jaw to the corner of the nare (the insertion depth can be marked with tape).
3. If the patient is awake and cooperative, the patient can be asked to breathe through each nostril while occluding the opposite nostril to determine which is more patent.
4. Prepare the selected nare by instilling several drops of OXYMETAZOLINE or PHENYLEPHRINE.
5. Lubricate the distal tip of the gastric tube with water soluble lubricant or 2% VISCOUS LIDOCAINE.
6. If the patient is awake, cooperative and able to sit upright, have them sit upright and flex their head forward (if not contraindicated).
7. Insert the tube into the selected nare and slowly advance the tube posteriorly parallel to the nasal canal (along the inferior aspect of the naris). The tube should not be directed upward as this may result in the tube being caught up in a blind recess at the middle turbinate.
8. The patient may gag as the tube approaches the larynx. If this occurs, temporarily halt advancement of the tube and instruct the patient to swallow. Advance the tube during swallowing as this will facilitate esophageal placement.
9. Advance the tube until gastric contents are returned or the desired depth of insertion is reached.
10. Confirm placement by auscultating for gastric sounds over the epigastrium while insufflating 20ml of air. Additionally, confirm placement by aspirating gastric contents.
11. Decompress the stomach of air and gastric contents by attaching the gastric tube to low suction or by manually aspirating with a large (Toomey) catheter tip syringe.
12. Secure the tube in place with tape.
13. Document gastric tube placement and confirmation by auscultation/aspiration of gastric contents. The volume of liquid output should also be documented.

E

Indication:
- Patient exhibiting behavior or actions that may be dangerous to the patient or medical providers.

Contraindications:
- None

Background:
Some patients may exhibit behavior or actions that may be dangerous to the patient or others. Any patient who may harm him or herself, or others, may be gently restrained to prevent injury to the patient or medical providers. Physical or chemical restraint must be humane and utilized only as a last resort. Other means to prevent injury to the patient or medical providers must be attempted first. These efforts may include reality orientation, distraction techniques, or other less restrictive therapeutic means.

Procedure:
1. Attempt less restrictive means of managing the patient.
2. Request law enforcement assistance.
3. Ensure there are sufficient personnel available to physically restrain the patient safely.
4. Restrain the patient in a lateral or supine position. No devices such as backboards, splints or other devices should be placed on top of the patient. Patients will never be restrained in the prone position.
5. The patient's upper extremities should be restrained with one arm at or above the level of the head and one arm at or below waist level if possible.
6. The restrained patient must be under constant observation by a licensed provider at all times. This includes ECG and SpO2 monitoring. Nasal waveform capnography may also be useful.
7. Extremities that are restrained will have a circulation check at least every 15 minutes. The first of these checks should occur as soon as possible after restraint application. This MUST be documented on the PCR.
8. Documentation on the PCR must include the reason for the use of restraints, the type of restraints utilized, and the time restraints were applied.
9. If the above actions are unsuccessful, or if the patient is resisting the restraints, chemical restraint should be utilized by advanced level providers in accordance with the _Behavioral Emergencies Protocol_ or the _Excited Delirium Protocol_ (chemical restraint may be considered earlier at the discretion of EMS providers).
10. If a patient is restrained by law enforcement personnel with handcuffs or other devices specific to law enforcement that EMS providers are unable to remove, a law enforcement officer must accompany the patient to the hospital in the transporting EMS vehicle or be immediately available.

Indication:
- Suspected pelvic fracture.
- Patient with blunt trauma (pedestrian struck, motorcycle crash, fall, and ejection) with hypotension or cardiac arrest.

Contraindications:
- Patient outside of appropriate body size/weight for specific device.

Background:
Pelvic fractures occurring from anterior-posterior (AP) compression (open book), vertical shear or a combination of both of these patterns may be associated with major life threatening bleeding. AP compression (open book) fractures commonly result from auto-pedestrian collisions, motorcycle crashes, direct crushing injuries to the pelvis, or falls from >12 feet. Vertical shear fractures occur when a high energy force is applied in a vertical plane and one half of the pelvis shifts upward. Bleeding associated with pelvic fractures may be venous, arterial, or a combination of both. Application of a pelvic binder provides stabilization of the pelvis and decreases the pelvic volume. The larger the pelvic volume, the greater the potential for hemorrhage. This protocol outlines the application of two commercially available pelvic binders, the SAM Sling™ and the T-POD®.

E

Procedure:

SAM Sling II™
1. Select appropriate size device for the patient's estimated waist circumference.
2. Remove objects from patient's pocket or pelvic area.
3. Place Sling black side up beneath patient at level of greater trochanters.
4. Place black strap through buckle and pull completely through.
5. Hold the orange strap and pull the black strap in opposite direction until you hear and feel the buckle click. Maintain tension and immediately press the black strap onto the surface of the sling to secure (you may hear a second click as the sling secures).
6. Once applied, the Sling should not be removed until the patient is in a definitive care setting and under the direction of a physician.

T-POD®
Patients under 50 lbs (23kg) may be too small to obtain the 6" gap needed for closure of the device.
1. Slide the belt under the supine patient under the pelvis.
2. Trim the belt, leaving a 6-8" gap over the center of the pelvis.
3. Apply the Velcro® backed pulley system to each side of the trimmed belt.
4. Slowly draw tension on the pull tab, creating simultaneous, circumferential compression.
5. Secure the Velcro® backed pull tab to the belt.
6. Once applied, the device should not be removed until the patient is in a definitive care setting and under the direction of a physician.

PEARLS:
- Pelvic fractures have a mortality rate of 5-50%, due mainly in part to the significant hemorrhage that may occur in the pelvis with minimal external signs.
- Delay in stabilization of the pelvis allows for continued hemorrhage.
- Excessive and repetitive manipulation ("rocking") the pelvis should be avoided. "Rocking" the pelvis is not an appropriate assessment technique in the patient with a suspected pelvic fracture.

Indication:
- Protocol directed medication administration via auto-injector.

Contraindications:
- None.

Background:
The auto-injector delivers a predetermined dose of medication via the intramuscular (IM) route. Use of an auto-injector is indicated as directed or recommended by protocol and when other administration routes are unavailable. The use of an auto-injector is highly recommended for the administration of some high risk medications like epinephrine. Risk of error is greatly decreased by the use of an auto-injector as there is no need to calculate or draw up a dose of medication. Medications commonly administered via auto-injector include epinephrine, atropine sulfate, naloxone and pralidoxime.

Procedure:
1. Check the label and expiration date on the auto-injector.
2. Confirm the Five "Rs": Right patient, Right medication, Right dose, Right route, and Right time.
3. If applicable, explain the procedure to the patient.
4. Locate the injection site (vastus lateralis located on the lateral aspect of the thigh. Injection is given at the mid-thigh level).
5. If time permits, expose the site and with a circular motion starting from the selected site outward, prepare the site with alcohol or chlorhexidine (auto-injectors are designed to work through clothing).
6. Remove the auto-injector from its storage container.
7. Form a fist around the auto-injector with the black tip facing down. Do NOT place your thumb over either end of the auto-injector.
8. Remove the safety cap from the auto-injector with your other hand.
9. Position the auto-injector at a 90° angle with the black or orange "needle end" cap against the desired injection site and press very firmly listening for an audible click indicating the needle has been deployed.
10. Hold the auto-injector in place for 10 seconds to allow for complete delivery of medication.
11. Remove the auto-injector and dispose of it properly.
12. Massage the injection site for 10 seconds to enhance medication delivery.
13. Observe the patient for response to the medication.
14. All patients receiving medication via auto-injector should be transported to the hospital for further evaluation and observation.

E

Indication:
- Patient requiring medications via the intramuscular (excludes auto-injector) or subcutaneous route.

Contraindications:
- None

Background:
The use intramuscular (IM) and subcutaneous (SC) injections are indicated for medications that are exclusively administered via these routes. Additionally, these routes provide an alternative to other routes (primarily the intravenous route) when other routes are unavailable. Due the rich blood supply to muscle, medications administered via the IM route are absorbed faster than those administered via the SC route. During shock states, the both the IM and SC routes are typically avoided as blood supply to both areas is reduced.

Procedure:

<u>Intramuscular</u>
1. Check the label, expiration date and appearance of the medication to be administered.
2. Confirm the Five "Rs": Right patient, Right medication, Right dose, Right route, and Right time.
3. If applicable, explain the procedure to the patient.
4. Select injection site appropriate for medication and volume to be administered. Use a 25g needle for aqueous medications and a 21-22g needle for oily or thicker medications. The length varies by site (see table 1 below).
5. Using anatomic landmarks, identify the selected injection site:
   - <u>Deltoid</u>
     - o Identify the bony portion of the shoulder where the clavicle and scapular meet [the acromioclavicular (AC) joint].
     - o Measure 2 fingers-width down from the AC joint.
   - <u>Vastus lateralis</u>
     - o Located on the anterior and lateral aspects of the thigh.
     - o Divide the area into thirds between the greater trochanter of the femur and the lateral femoral condyle. Injection site is the middle third.
   - <u>Ventrogluteal</u>
     - o Place heel of your palm of your hand on the patient's greater trochanter of the femur.
     - o Place your index finger on the anterior superior iliac spine and spread your other fingers posteriorly.
     - o Injection site is in the V formed between the index finger and the second finger.
7. With a circular motion starting from the selected site outward, prepare the site with alcohol or chlorhexidine.
8. With one hand, stretch or flatten the skin overlying the selected site.
9. In the other hand, hold the syringe like a dart and quickly insert the needle into the tissue and muscle at a 90 degree angle.
10. After the medication is injected, quickly withdraw the syringe and dispose of it properly.
11. Gently massage over the injection site to increase medication absorption.
12. Apply firm pressure over the site and apply an adhesive bandage.

Table 1: IM Injection Needle Length and Medication Volumes

Deltoid IM injection site:

Site	Needle Length	Medication Volume
Posterior Deltoid	Length 0.5 to 1"	Up to 2ml, contingent upon muscle mass development.
Vastus Lateralis	Length 1.0 to 1.5"	Up to 2ml in adults and children.
Ventrogluteal	Length 1.0 to 1.5"	2-5 ml in the adult, up to 2 ml in children.

E

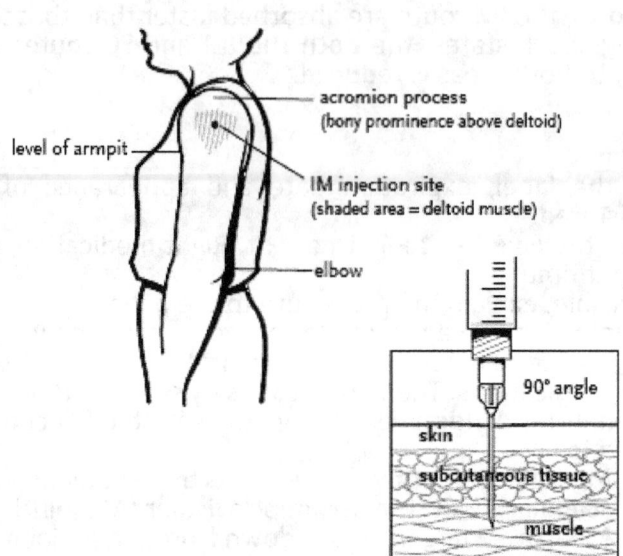

P

Subcutaneous (SC)
1. Check the label, expiration date and appearance of the medication to be administered.
2. Confirm the Five "Rs": Right patient, Right medication, Right dose, Right route, and Right time.
3. If applicable, explain the procedure to the patient.
4. Identify the desired injection site. Common SC injection sites include the upper outer triceps area, the outer aspect of the upper thigh, and the upper buttocks.
5. With a circular motion starting from the selected site outward, prepare the site with alcohol or chlorhexidine.
6. Without contaminating the injection site, raise a fold of skin between your thumb and forefinger and insert the needle at a 45-90° angle.
7. After the medication is injected, quickly withdraw the syringe and dispose of it properly.
8. Apply firm pressure over the site and apply an adhesive bandage.

PEARLS:
- When administering IM or SC injections, aspirating to exclude inadvertent vascular placement of the needle is no longer recommended.

E

Indication:
- Life threatening hemorrhage that cannot be controlled by other means.
- Serious or life threatening extremity hemorrhage where conditions (patient location, tactical or hazmat environment) prevent the use of standard hemorrhage control techniques.
- Life threatening condition(s) that require immediate attention and significant extremity hemorrhage where the use of a tourniquet is more expedient than standard hemorrhage control.

Contraindications:
- None

Background:
Tourniquets have a long history in emergency care. In the recent past, it was taught that they should only be used as a "last resort". However, due to increasing data from the battlefield, the threshold for tourniquet application should be very low. The mainstays for the control of hemorrhage from the extremities are direct pressure and tourniquet application. The use of elevation and pressure points are no longer recommended. While the Rhode department of Health Center for Emergency Medical Services does not endorse any particular brand of hemostatic tourniquet, is recommended that only tourniquets that have been evaluated and approved by the Committee on Tactical Combat Casualty Care (CoTCCC) be used.

Procedure:
Extremity Tourniquet Application
1. Place the tourniquet proximal to the wound.
2. Tighten the tourniquet until hemorrhage stops and there is loss of the distal pulse.
3. Secure the tourniquet(s). The tourniquet should be easily visible on the limb and should not be covered.
4. Note the time of tourniquet application.
5. If hemorrhage control is not achieved with the application of the 1st tourniquet, apply a 2nd tourniquet just distal to the first one (the first one should be left in place).
6. Do not remove a tourniquet once hemostasis has been achieved.
7. Provide wound care as per the *Wound Care Procedure Protocol*.
8. Provide analgesia as indicated as per age appropriate *Patient Comfort Protocol*.

Junctional Tourniquets
If available and providers are appropriately familiar with device operation and application, a junctional tourniquet may be used for junctional hemorrhage control.

PEARLS:
- Delay in placement of a tourniquet for life threatening hemorrhage significantly increases mortality. Do not wait for hemodynamic compromise to apply a tourniquet.
- Tourniquets should not be applied directly over the knee or elbow. If a wound is just distal to a joint, place the tourniquet just proximal to the joint. Additionally, do not put a tourniquets directly over a holster or a cargo pocket that contains bulky items.
- Tourniquets should not be loosened to allow blood flow to return to the injured extremity.
- Damage to the limb from tourniquet application is unlikely if it is removed in several hours.
- It is to be expected that the patient will experience pain in the affected extremity after tourniquet application.
- Currently CoTCCC tourniquets include the C-A-T (Combat Application Tourniquet) and the SOFTT (SOF Tactical Tourniquet).

E

Indication:
- Life threatening compressible hemorrhage that is not amenable to tourniquet application.

Contraindications:
- None

Background:
Hemostatic dressing offer an alternative for the control of hemorrhage when it is not amenable to tourniquet application. Hemostatic dressings are impregnated with material (kalolin or chitosan) that are pro-coagulant (enhance clotting). There are three hemostatic dressings recognized by the Committee on Tactical Combat Casualty Care (Co-TCC). These in include Quick Clot Combat Gauze©, Celox™ Gauze, and ChitoGauze®. Current T-CCC guidelines preferentially recommend Quick Clot Combat Gauze©. However, the guidelines allow for alternative use of Celox™ gauze and ChitoGauze® in the event Combat Gauze™ is not available. Hemostatic dressings are impregnated with materials that are pro-coagulant (enhance clotting). Hemostatic dressings must be deeply packed into the wound and direct pressure must be maintained for 3 minutes following application. Providers should be trained in and practice the application technique for hemostatic dressings.

Procedure:
1. Expose the injury by opening or cutting away the patients clothing.
2. If possible, remove excess blood from the wound while preserving any clots that may have formed.
3. Locate the source of the most active bleeding.
4. Remove the hemostatic dressing from its sterile package and pack it tightly into the wound directly over the site of the most active bleeding (more than one dressing may be required to achieve hemorrhage control).
5. Apply direct pressure quickly with enough force to stop the bleeding, hold direct pressure for a minimum of 3 minutes.
6. Reassess for bleeding control.
7. More dressing may be packed into the wound as necessary to stop any continued bleeding.
8. Secure the hemostatic dressing in place with a compression bandage (Ace wrap, roller gauze, emergency trauma bandage).

PEARLS:
- Training resources and videos are available from the manufactures of the above mentioned hemostatic dressings as follows:

Combat Gauze
http://www.z-medica.com/military/Home.aspx
Celox Gauze
http://www.celoxmedical.com/usa/products/celox-gauze/
ChitoGauze
http://www.hemcon.com/Products/ChitoGauzeHemostaticGauzeOverview.aspx/

Indication:
- Protection and care of open wounds prior to and during transportation.

Contraindications:
- None

Background:
The goals of prehospital wound care include control of related hemorrhage and minimizing the risk for infection.

Procedure:
1. Use appropriate personal protective equipment, including gloves, gown, eye protection and mask as indicated.
2. If active hemorrhage is occurring, control it following the *External Hemorrhage Control Protocol.*
3. Once hemorrhage is controlled, irrigate contaminated wounds with sterile normal saline solution as appropriate. Avoid irrigation if hemorrhage is difficult to control.
4. Cover wounds with a sterile dressing and secure with a bandage or compression wrap. Check distal pulses, sensation, and motor function to ensure the bandage/wrap is not too tight.
5. Monitor wounds and or dressings throughout transport for bleeding.
6. If able, determine the patient's tetanus immunization status.
7. Document the wound(s) and care in the PCR.

E

E

Indication:
- Patient requiring rapid medication administration when other routes are not immediately available.

Contraindications:
- None

Background:
Intranasal medication delivery provides for the rapid administration of certain medications (naloxone, midazolam, fentanyl) when other routes are unavailable or unsafe.

Procedure:
1. Draw up appropriate dose of medication into syringe (1 ml max).
2. Confirm the Five Rs: Right patient, Right medication, Right dose, Right route, and Right time.
3. Attach a mucosal atomization device to the syringe.
4. Using the free hand to hold the occiput of the head stable, place the tip of the mucosal atomization device snugly against the nostril aiming slightly up and outward (toward the top of the ear).
5. Briskly compress the syringe plunger. Then deliver half of the medication into the nostril.
6. Dispose of the syringe and atomizing device in an approved sharps container.

PEARLS:
- Minimize the volume, and maximize concentration of medication (1/3 mL per nostril is ideal, 1 mL is max).
- Maximize the total mucosal absorptive surface area. Atomize the medication (rather than dripping it in). Use both nostrils to double the absorptive surface area. Aim slightly up and outwards to cover the turbinates and olfactory mucosa.
- Beware of abnormal mucosal characteristics. Mucous, blood and vasoconstrictors reduce absorption. Suction the nostrils or consider alternate drug delivery method in these situations.

Indication:
- Significant epistaxis unamenable to direct pressure and/or a vasoconstrictor.

Contraindications:
- None.

Background:
Placement of an intranasal tampon (packing) allows for the application of constant local pressure to the nasal septum. Nasal packing works by the application of local direct pressure, reduction of mucosal irritation (which decreases bleeding) and clot formation surrounding the foreign body, which enhances pressure. This protocol outlines the use of the Rapid Rhino® nasal tampon. Other similar devices may be utilized with service Medical Director approval.

Procedure:

Rapid Rhino®
1.   Select the appropriate size/type (anterior/posterior) device.
2.   Soak the nasal tampon in sterile water for a FULL 30 seconds.
3.   Insert the device along the superior aspect of the hard palate until the blue indicator is   past the nares.
4.   Utilizing a 20ml syringe, inflate the device with air only. Monitor the pilot cuff for direct tactile feedback; stop inflation when the pilot cuff becomes rounded and feels firm when squeezed (the cuff should be inflated to provide gentle, low pressure delivering the fabric of the device directly to the bleeding site.
5.   Reassess after 15-20 minutes; re-inflate to ensure proper pressure (if necessary) and tape to it to the patient's cheek away from the upper lip.
6.   Patients with nasal packing placed in the field will require antibiotics and therefore must be transported to a Hospital Emergency Facility.

P

PEARLS:
- Anterior epistaxis is identified by blood draining primarily from one or both nare (90% of cases are anterior in etiology).
- Posterior epistaxis is identified by the observation of blood draining into the posterior pharynx.
- Prolonged nasal packing has been associated with toxic shock syndrome.

Indication:
- Life threatening hemorrhage from wounds in the groin or axilla not amenable to tourniquet application.

Contraindications:
- None.

Background:
XSTAT is a hemostatic dressing for the control of severe, life-threatening bleeding from junctional wounds in the groin or axilla not amenable to tourniquet application in adults and adolescents. XSTAT contains rapidly expanding cellulose sponges coated with Chitosan. These sponges expand and create a barrier to blood flow, present a large surface area for clotting, and provide gentle pressure. XSTAT is a temporary dressing for use up to four (4) hours until surgical intervention can be accomplished. XSTAT should only be used for patients at high risk for immediate life-threatening bleeding from, hemodynamically significant, non-compressible junctional wounds. XSTAT is not indicated for use in the thorax, the pleural cavity, the mediastinum, the abdomen, the retroperitoneal space, the sacral space above the inguinal ligament, or tissues above the clavicle.

E

Procedure:
1. Expose the injury by opening or cutting away the patient's clothing.
2. Open the package and remove applicator.
3. Pull the black handle out and away from barrel until it stops and locks.
4. Place the tip of the applicator into the wound track as close to the bleeding source as possible.
5. Firmly depress black handle to deploy sponges into the wound. Material should flow freely into the wound. Deploy XSTAT within 30 seconds of insertion into wound. DO NOT attempt to forcefully eject the material from the applicator. If resistance is met, pull back slightly on the applicator body to create additional packing space, then continue to depress handle.
6. Use additional applicators as necessary to completely pack the wound with sponges. DO NOT attempt to remove sponges from wound. Sponges must be removed intraoperatively with the capability and equipment for achieving proximal and distal vascular control.
7. Cover the wound with an occlusive or pressure dressing (if available, use an elastic bandage).
8. If bleeding persists, apply manual pressure until bleeding is controlled.
9. Note and document the time of application. Assess the patient for peripheral circulation and document presence of distal pulse and the time of application.

PEARLS:
- Vascular compression greater than four hours is not recommended due to concerns related to limb ischemia.
- Triangular segments of the applicator tip may break away from applicator during treatment. If this occurs, do not attempts to retrieve it from the wound. Document the occurrence and communicate this information to receiving facility staff.
- A XSTAT training video and training information may be found at www.revmedx.com.

Indication:
- Patient with moderate to severe pain.

Contraindications:
- Altered mental status
- Acute intoxication or drug use
- Pregnancy (except during delivery)
- Blunt or penetrating chest trauma/pneumothorax
- Craniofacial injury/traumatic brain injury/increased intracranial pressure
- Undifferentiated abdominal pain
- Diving emergencies (decompression illness)
- Respiratory distress
- Maxillofacial abnormalities/facial trauma or burns
- Status-post retinal surgery

Background:

N_2O has analgesic effects and is useful in the treatment of mild to moderate pain, or as a bridge to IV analgesia. The exact mechanism of action for N_2O is unknown, but its effects take place within the pain centers of the brain and spinal cord. It is thought to affect the release of endogenous neurotransmitters such as opioid peptides and serotonin. The release of these neurotransmitters is thought to activate descending pain pathways that inhibit pain transmission. Additionally, it is thought to have an effect on the gamma aminobutyric acid (GABA) receptors increasing inhibition of nerve cells causing drowsiness and sleep. The onset of action is 30-60 seconds and the peak effect is seen within 2-5 minutes. N_2O is able to diffuse from the blood in to closed gas spaces (i.e. bowel, middle ear, pneumothorax) causing them to expand. Because N_2O is self-administered by the patient, the patient must be able to understand and follow directions. N_2O is blended with oxygen and in the EMS setting is delivered in a 50/50 concentration.

E

Procedure:
1. Have the patient (not the provider) hold the mask tightly against the patient's face.
2. Instruct the patient not to talk or remove the mask between breaths.
3. Instruct the patient to breath normally.
4. Monitor the patient's vital signs and SpO_2.
5. Continuously assess the patient for lightheadedness, restlessness, and nausea.
6. Document administration of nitrous oxide 50/50% on the PCR.
7. If bleeding persists, apply manual pressure until bleeding is controlled.
8. Note and document the time of application. Assess the patient for peripheral circulation and document presence of distal pulse and the time of application.

A C P

Indication:
- Patient requiring medications or fluids via the intravenous route.
- Potentially unstable patient requiring precautionary intravenous access.

Contraindications:
- None

Background:
Peripheral intravenous access may be established any patient requiring medications or fluids via the intravenous route as directed by protocol(s). Precautionary peripheral intravenous access may also be established in in potentially unstable patients. A Saline Lock may be utilized as an alternative to the use of an intravenous administration set and intravenous fluid bag in every protocol at the discretion of the EMS provider.

Procedure:

Peripheral Intravenous Insertion
1. Explain procedure to patient, if applicable.
2. Select appropriate intravenous fluid.
3. Select access site appropriate for the patient's condition:
   - Upper or lower extremities
   - Scalp vein [infant] (Paramedic only)
   - External jugular vein (Paramedic only)
4. Apply a venous constricting band (if appropriate to anatomic site) proximal to the intended insertion site.
5. Cleanse the site with chlorhexidine or alcohol (for infants under 2 months, use alcohol 70%).
6. Insert the needle with the bevel up into the skin in a steady, deliberate motion until a flashback of blood is observed in the flash chamber of the device.
7. Advance the catheter into the vein (never reinsert the needle into the catheter).
8. Actuate the needle retraction mechanism on the device and dispose of the device in an approved sharps container.
9. Obtain blood samples if applicable.
10. If able, use your thumb or a finger to occlude the catheterized vein proximal to the catheter. Remove the tourniquet and attach the intravenous administration set to the catheter.
11. Open the flow control on the administration set and assure the free flow of IV fluid and then adjust the flow rate as indicated.
12. Cover the site with a sterile dressing and secure the catheter and tubing.
13. Label the site with date, time, catheter gauge, and initials of the person who started the IV.
14. Document the procedure on the Patient Care Report. Documentation should include the time the IV was started, the gauge of the catheter, and the identification of the provider who started the IV.

Saline Lock or Intermittent Needle Therapy (INT) Device
1. Prepare saline lock and extension set by flushing with 2mL of normal saline to ensure the line and lock are air-free.
2. Perform steps 1-9 (as applicable) for peripheral intravenous access.
3. If able, use your thumb or a finger to occlude the catheterized vein proximal to the catheter.
4. Attach the saline Lock extension set to catheter (be sure the luer lock is secure).
5. Cover the site with a sterile dressing and secure the catheter (be sure to leave the connection between the IV tubing and the catheter hub uncovered)
6. Label the site with date, time, catheter gauge, and initials of the person who started the IV.
7. Document the procedure on the Patient Care Report. Documentation should include the time the IV was started, the gauge of the catheter, and the identification of the provider who started the IV.

PEARLS:

- Do not place IVs (or take a blood pressure) in and extremely with an AV fistula or the same side as a mastectomy.

A
C
P

Indication:
- Patient requiring rapid IV access for medication or fluid administration when regular peripheral IV access is unavailable or secondary access is required in a critically ill or injured patient.
- Primary vascular access in cardiac arrest (when placed superior to the diaphragm).

Contraindications:
- Fracture or previous IO access attempt (within 24 hours) in target bone proximal to proposed access site.
- Infection or burn over proposed access site.
- Prosthetic joint at or proximal to the proposed access site.
- Osteogenesis imperfecta.

Background:
Intraosseous access allows for expedient and functional vascular access for the administration of medications and fluids. Intraosseous access in indicated in patients requiring rapid IV access for medication or fluid administration when regular peripheral IV access is unavailable or secondary access is required in a critically ill or injured patient. In cardiac arrest, when placed superior to the diaphragm, IO may be used as the primary vascular access. All medications contained in these protocols that are routinely administered via the conventional intravenous route can be administered via the intraosseous route. EMS providers may utilize any intraosseous vascular access device that is approved by the FDA with the exception of spring loaded devices for sternal placement,. The validation and documentation of the individual EMS provider's competency utilizing a particular device is ultimately the responsibility of the service Medical Director. EMS providers should follow the device manufactures recommendations for use and locations for anatomic placement.

Procedure:
1. If possible, place the patient in a supine position.
2. If applicable, explain the procedure to the patient.
3. Select appropriate site as per specific device FDA approved insertion site(s). These may include the proximal or distal tibia, the proximal humerus or the sternum.
4. Identify bony and other anatomic landmarks specific to the selected insertion site.
5. Prepare the site with alcohol or chlorhexidine.
6. Insert the device manufacturer's recommended procedure.
7. Confirm proper placement of the device following the device manufacturer's recommendation. Generally, devices are properly placed if they stand without support and infused fluids do not infiltrate at the site. While the aspiration of bone marrow is confirmatory, the absence of bone marrow aspiration is inconclusive.
8. If the patient is awake and alert, provide intramedullary anesthesia by slowly injecting 2% (cardiac) LIDOCAINE through the IO device. The dose for the adult patient is 20-50 mg and the dose for the pediatric patient is 0.5 mg/kg.
9. Flush the device with 10 ml of normal saline prior to use.
10. Stabilize the device utilizing a commercially available stabilization device or stabilize the needle on both sides with sterile dressings and secure with tape avoiding tension on the needle. Additional manual stabilization is recommended during resuscitation involving infants or small pediatric patients.
11. Following the administration of medication, flush the device with 10 ml of IV fluid. For the continuous infusion of resuscitation fluids, a pressure infuser bag should be used. Vasopressors should be administered via an IV infusion pump.
12. Document procedure, including size and type of device used, number of attempts, anatomic location, +/- use of anesthetic, and any complications.

PEARLS:
- Intraosseous placement in cardiac arrest should not interfere with or interrupt chest compressions.

Indication:
- Patient with a life threatening condition or in extremis from a readily treatable etiology (i.e. pulmonary edema) requiring venous access when traditional means are unsuccessful/ unavailable.

Contraindications:
None.

Background:
Some patients requiring long term care may have alternate vascular access devices placed. These devices include centrally placed venous lines, peripherally inserted central line catheters (PICC line), internal subcutaneous infusion ports (Portacath), hemodialysis AV fistulas and grafts, or tunneled catheters (Broviac, Hickman, Groshon, Leonard etc.). In the event a patient with one of these devices is experiencing a life threatening condition or is in extremis from a readily treatable etiology (i.e. pulmonary edema) and venous access through traditional means is unsuccessful/unavailable, paramedics may access these devices.

Procedure:

P

Centrally Placed Venous Catheter (single or multi-lumen)
1. Select a port to access. Fluids and medications administered via any of the ports will end up in the same place (central venous circulation), so any available port can be used for emergent fluid or medication administration. If there is a port reserved for total parental nutrition (TPN) only, it will usually be labeled as such and should be avoided if possible.
2. Utilize good hand hygiene with either an alcohol based cleanser or soap and water.
3. Cleanse the port with chlorohexidine.
4. Using sterile technique, access the port with a 10 ml syringe and aspirate 5-10 ml of blood. Detach the syringe and attach another syringe with 5-10 ml of normal saline and gently flush the port. If there is resistance, evidence of infiltration, or any concern that the catheter may be clotted or dislodged, do not use the catheter.
5. Remove the syringe, attach an IV administration set and proceed as normal, opening the line and insuring patency.

Peripherally Inserted Central Catheters (PICC)
PICC lines are usually inserted into the distal superior vena cava or the right atrium via the antecubital vein.
1. Utilize good hand hygiene with either an alcohol based cleanser or soap and water.
2. Select a port on one of the catheters (if two sizes are available, select the larger of the two).
3. Cleanse the port with chlorohexidine.
4. Attach a 10 ml syringe with 5ml of normal saline drawn up into it to the selected port's needleless connector.
5. Unclamp the selected catheter lumen (if applicable) and inject the normal saline into the port and then aspirate to achieve blood return, indicating successful access (if resistance is met, repeat on another port, if a port without resistance is not identified, do not use the device).
6. Remove the syringe, attach an IV administration set and proceed as normal, opening the line and insuring patency.

Tunneled Catheters (Broviac, Hickman, Groshong, Leonard)
Tunneled catheters are also inserted into the distal superior vena cava or the right atrium via the cephalic vein. These catheters enter the skin through an incision on the chest. Most of these devices are heparinized and protected with an injectable cap.

1. Utilize good hand hygiene with either an alcohol based cleanser or soap and water.
2. Select an appropriate port for access. If more than one is available, select the largest one (if they are color coded, select the blue or brown one).
3. Clamp off the selected catheter. Expose the end of the hub (it may have a cap or be taped over) and cleanse it with chlorhexidine.
4. Attach a 10 ml syringe to the selected port's needleless connector.
5. With the syringe attached, unclamp the catheter and aspirate 5ml of blood and heparin (there should be no resistance to aspiration, if unable to aspirate, do not use the catheter).
6. Detach the syringe and dispose of it properly in an approved sharps container.
7. Without contaminating the needless connector, attach a second syringe with 10 ml of normal saline and flush the catheter (the catheter should flush freely, if it does not, do not use the catheter).
8. Remove the syringe, attach an IV administration set and proceed as normal, opening the line and insuring patency.

Internal Subcutaneous Infusion Ports (Portacath)
1. Utilize good hand hygiene with either an alcohol based cleanser or soap and water.
2. Locate the port by visualization and palpation (subclavicular, dome shaped prominence).
3. Vigorously cleanse the site with chlorhexidine and allow the site to air dry.
4. Attach a 10 ml syringe prefilled with normal saline to the infusion port cap of the extension tubing of a non-coring, right angle (Huber) needle and prime the tubing and needle.
5. Don a surgical mask and mask the patient, don sterile surgical gloves (if available).
6. Secure the access point port firmly between two fingers and firmly insert the non-coring needle into the port, entering at a 90° angle.
7. Aspirate 3-5 ml of blood with the syringe (if unable to aspirate blood, do not use the port), detach the syringe and dispose of the syringe and aspirated blood properly in an approved sharps container.
8. Attach a 10 ml syringe prefilled with normal saline to the infusion port cap of the extension tubing of the non-coring needle. Flush the device with 3-5 ml of normal saline (if the device does not flush easily, do not use the device).
9. Remove the syringe, attach an IV administration set and proceed as normal, opening the line and insuring patency.
10. Cover the needle and insertion site with a sterile occlusive dressing.

AV Fistula or Graft (cardiac arrest and periarrest patients only)
1. In the patient with spontaneous circulation, palpate the fistula for a thrill (ie, a vibration felt over the fistula during systole, which indicates proper functioning) and document its presence.
2. Apply a venous tourniquet proximal to the fistula site.
3. Cleanse the site with chlorhexidine.
4. Insert a 16-18g catheter-over-the needle device, needle or butterfly needle into the fistula at a 20-35° angle (if accessing a graft, enter at a 45° angle). Insert until blood return is observed. Once blood return is observed, insert the catheter or needle to the hub.
5. Remove the venous tourniquet (if utilized).
6. Secure the catheter or needle in place.
7. Attach an IV administration set and proceed as normal, opening the line and insuring patency (blood flows through fistulas at high velocity and fluids may need to administered under pressure).
8. Cover the needle and insertion site with a sterile occlusive dressing.

PEARLS:

- PICC lines will not tolerate rapid infusions or infusions under pressure.
- Avoid measuring a blood pressure in the same arm with a PICC inserted.
- Do not exceed recommended flow rates.

Device	Size	Max Flow Rate
PICC	< 2.0 fr	125 ml/hr
PICC	> 2.0 fr	250 ml/hr
Groshong PICC	3 fr	240 ml/hr
Grsohong PICC NXT	4 fr	540 ml/hr
Groshong PICC NXT	5 fr	200 ml/hr
Hickman/Broviac (Power Port)	8-9.5 fr	3000 ml/hr

- AV fistulas are created surgically by creating an anastomosis between an artery and a vein. An AV fistula can usually be differentiated from an AV graft by the presence of two scars.
- An AV graft is created surgically by connecting an artery and a vein together with a synthetic graft. An AV graft usually has only one scar associated with it.
- The Groshong catheter is very similar to the Hickman catheter, but has a valve at the tip of the catheter which makes it unnecessary to leave a high concentration of heparin in the catheter.

P

Indication:
- Primary venous access in a neonate ≤ ~ 1 week of age requiring resuscitation or emergent medication administration.

Contraindications:
- Peritonitis
- Omphalitis
- Necrotizing enterocolitis

Background:
Cannulation of the umbilical vein provides vascular access in the neonate ≤ ~ 1 week of age requiring resuscitation or emergent medication administration. Cannulation of the umbilical vein should be reserved for the unstable patient in which peripheral or intraosseous access is unavailable.

Procedure:
1. Double wrap a piece of umbilical tape around the base of the cord.
3. Utilize sterile technique (sterile gloves, mask, drapes) as much is possible.
3. Prepare the cord stump and surrounding abdomen with betadine solution.
4. If the stump is not fresh, cut the stump horizontally with a #15 scalpel approximately 1.5-2 cm from the abdominal wall.
5. If there is bleeding, hemostasis can be achieved by applying gentle tension on the umbilical tape.
6. Identify the umbilical vasculature. The two arteries are the thick-walled small vessels and the one thin-walled vessel is the vein. The umbilical vein (UV) is the largest of the three vessels and is usually located in the 12-o'clock position.
7. If present, utilizing forceps, gently remove and clots from the UV.
8. Utilizing forceps, open and dilate the vein (the vein is usually open and will not require dilation).
9. Insert an appropriately sized UV catheter (3.5 fr for premature infants and 5fr for full-term infants). The catheter should be inserted just to the point that blood returns (low line placement).
10. Secure the catheter in place.

PEARLS:

- Advancing the catheter beyond the point at which blood is returned may result in placement of the catheter in the portal vein.
- Portal vein placement should be suspected if resistance to advancement is met. If portal vein placement is suspected, withdraw the catheter to the appropriate point.

Section 8:   Special Operations

08.00   Fire Ground and Extended Operations Responder Rehabilitation

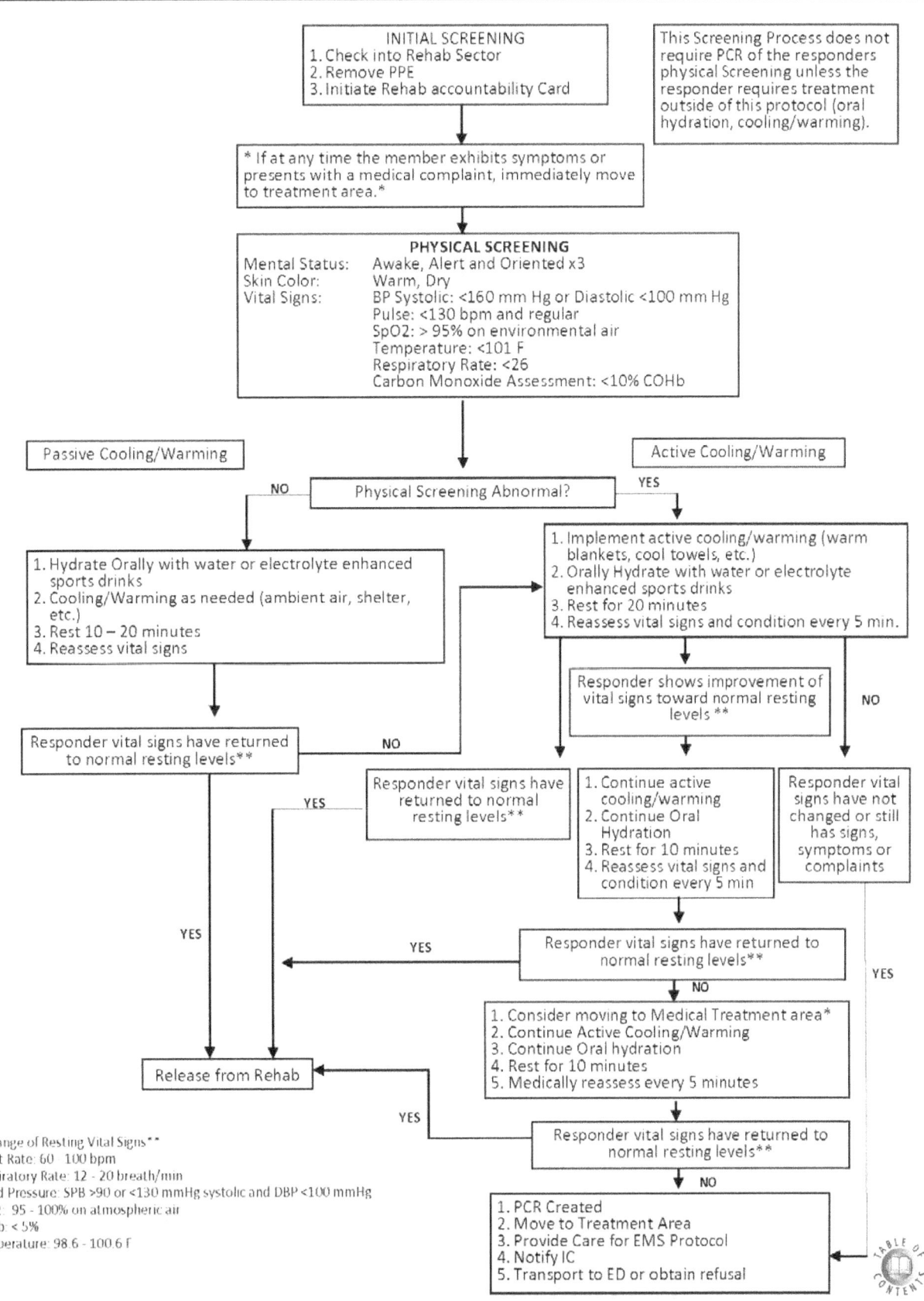

INITIAL SCREENING
1. Check into Rehab Sector
2. Remove PPE
3. Initiate Rehab accountability Card

This Screening Process does not require PCR of the responders physical Screening unless the responder requires treatment outside of this protocol (oral hydration, cooling/warming).

* If at any time the member exhibits symptoms or presents with a medical complaint, immediately move to treatment area.*

PHYSICAL SCREENING
Mental Status: Awake, Alert and Oriented x3
Skin Color: Warm, Dry
Vital Signs: BP Systolic: <160 mm Hg or Diastolic <100 mm Hg
   Pulse: <130 bpm and regular
   SpO2: > 95% on environmental air
   Temperature: <101 F
   Respiratory Rate: <26
   Carbon Monoxide Assessment: <10% COHb

Passive Cooling/Warming

Active Cooling/Warming

Physical Screening Abnormal?   NO   YES

1. Hydrate Orally with water or electrolyte enhanced sports drinks
2. Cooling/Warming as needed (ambient air, shelter, etc.)
3. Rest 10 – 20 minutes
4. Reassess vital signs

1. Implement active cooling/warming (warm blankets, cool towels, etc.)
2. Orally Hydrate with water or electrolyte enhanced sports drinks
3. Rest for 20 minutes
4. Reassess vital signs and condition every 5 min.

Responder shows improvement of vital signs toward normal resting levels **   NO

Responder vital signs have returned to normal resting levels**   NO

Responder vital signs have returned to normal resting levels**   YES

1. Continue active cooling/warming
2. Continue Oral Hydration
3. Rest for 10 minutes
4. Reassess vital signs and condition every 5 min

Responder vital signs have not changed or still has signs, symptoms or complaints

Responder vital signs have returned to normal resting levels**   YES   NO   YES

YES

1. Consider moving to Medical Treatment area*
2. Continue Active Cooling/Warming
3. Continue Oral hydration
4. Rest for 10 minutes
5. Medically reassess every 5 minutes

Release from Rehab

YES

Responder vital signs have returned to normal resting levels**   NO

** Range of Resting Vital Signs**
Heart Rate: 60 - 100 bpm
Respiratory Rate: 12 - 20 breath/min
Blood Pressure: SPB >90 or <130 mmHg systolic and DBP <100 mmHg
SpO2: 95 - 100% on atmospheric air
COHb: < 5%
Temperature: 98.6 - 100.6 F

1. PCR Created
2. Move to Treatment Area
3. Provide Care for EMS Protocol
4. Notify IC
5. Transport to ED or obtain refusal

TABLE OF CONTENTS

PEARLS:

- Remove PPE, body armor, chemical suits, SCBA, turnout gear and other equipment as indicated.
- Firefighters should consume at least 8 oz of fluid between SCBA change out.
- Responders taking antihistamines, antihypertensive agents, diuretics or stimulants are at increased risk for cold and heat stress.
- If using an electrolyte sports drink (ie. Gatorade) for oral hydration, it should be diluted 50/50 with water.

Section 9:   APPENDIX

Medication and Formulation	Minimum Quantity	E	A/C	P
Acetaminophen (oral)	Four (4) 500 mg tabs	R	R	R
Acetaminophen (oral suspension)	One (1) 20 ml bottle (160 mg/5 ml)	R	R	R
Acetaminophen (rectal)	Three (3) 325 mg or five (5) 120 mg suppositories	R	R	R
Activated charcoal (oral)	Two (2) 25 gm unit dose or one (1) 50g unit dose	R	R	R
Adenosine (injectable)	24 mg (prefilled syringe/vial)	X	R	R
Albuterol 0.083% (inhalation)	10 mg (2.5 mg unit dose bullets)	R	R	R
Amiodarone (injectable)	450 mg (prefilled syringe/vial/ampule)	X	R	R
Amiodarone (infusion)	360 mg/200 ml (premixed bag)	X	X	R
Aspirin (oral)	Twenty (20) 81 mg tabs	R	R	R
Aspirin (oral)	Four (4) 325 mg tabs	O	O	O
Atropine sulfate (injectable)	3 mg (prefilled syringe)	X	R	R
Blood products and universal products		X	X	TM
Calcium chloride 10% (injectable)	2 gm (prefilled syringe/vial)	X	R/C	R/C
Calcium gluconate 10% (injectable)	6 gm (vial/ampule)	X	R/C	R/C
Calcium gluconate 2.5% gel (topical)	Service optional/if available	O	O	O
Calcium gluconate 2.5% (inhalation)	Service optional/if available	X	X	O
Cefazolin (injectable)	Two (2) 1 gm (premixed formulation/vial/ampule)	X	X	R
Dexamethasone (injectable)	10 mg (vial/ampule)	X	C	R
Dextrose 5% (injectable)	One (1) 250 ml bag, One (1) 500 ml bag	X	X	R
Dextrose 10% (injectable)	Two (2) 250 ml bags	X	R	R
Dextrose 25% (injectable)	2.5 gm (prefilled syringe/vial)	X	O	O
Dextrose 50% (injectable)	25 gm (prefilled syringe/vial)	X	R	R
Diazepam (injectable)	20 mg (prefilled syringe/vial/ampule)	X	O/C	O/C
Diltiazem (injectable)	50 mg (prefilled syringe/vial)	X	R	R
Diltiazem (for infusion)	100 mg (vial or other)	X	X	O
Diphenhydramine (injectable)	50 mg (vial/ampule)	X	R	R
Diphenhydramine (oral)	50 mg (tabs/caps/powder/elixir)	X	R	R
Dopamine HCL (infusion)	400 mg/250 ml or 800 mg/500 ml (premixed bag)	X	X	R
Droperidol (injectable)	20 mg (vial/ampule)	X	X	C
DuoDote Antidote Kit	Service optional/if available	O/C	O/C	O/C
Enalaprilat (injectable)	1.25 mg (vial/ampule)	X	X	R

Medication and Formulation	Minimum Quantity	E	A/C	P
Epinephrine 1:10,000 (injectable)	8 mg (prefilled syringe)	X	R	R
Epinephrine 1:1,000 (injectable)	3 mg (auto-injector/prefilled syringe/vial/ampule)	R	R	R
Epinephrine 2.25% (inhalation)	2 (two) 0.5ml (unit dose bullet)	X	R	R
Etomidate	40 mg (any combination of vials)	X	X	O
Famotidine (oral)	Two (2) 20 mg tabs	X	X	R
Famotidine (injectable)	40 mg (vial/ampule)	X	X	R
Fentanyl (injectable)	300 mcg (vial/ampule)	X	R	R
Furosemide (injectable)	80 mg (prefilled syringe/vial/ampule)	X	R	R
Glucagon (injectable)	1 mg (vial)	R	R	R
Glucose (oral)	Two (2) 15 gm (gel)	R	R	R
Haloperidol (injectable)	20 mg (vial/ampule)	X	X	R
Hydrocortisone (injectable)	100 mg (vial)	X	R	R
Hydroxocobalamin (infusion)	Service optional/if available	X	O	O
Ibuprofen (oral)	Eight (8) 200 mg tabs	R	R	R
Ibuprofen (oral suspension)	One (1) 20 ml bottle (100 mg/5 ml)	R	R	R
20% Fat Emulsion (infusion)	250 ml (bottle)	X	X	R
Ipratropium bromide (inhalation)	0.5 mg unit dose or may be combined with albuterol in a unit dose bullet.	O	R	R
Ketamine (injectable)	One (1) 10 ml (50 mg/ml) **and** one (1)20 ml (10 mg/ml) ampule/vial.	X	X	R
Ketorolac (injectable)	60 mg (prefilled syringe/ampule)	X	X	R
Lactated Ringer's (injectable)	Three (3) 1000 ml bags	X	R	R
Labetalol *	N/A	X	X	TM
Levalbuterol (inhalation)	1.25 mg/3ml (unit dose/MDI)	X	X	O
Levetiracetam	4 (four) 500 mg vials	X	X	O
Lidocaine (injectable)	300 mg (prefilled syringe)	X	R	R
Lidocaine (IV infusion)	1 gm/250 ml or 2 gm/500 ml (premixed)	X	X	R
Lorazepam (injectable)	10 mg (prefilled syringe/vial)	X	O/C	O/C
Magnesium Sulfate (injectable)	4 gm (premixed/vial)	X	X	R
Mark 1 NAAK	N/A	O/C	O/C	O/C
Methylprednisolone (injectable)	125 mg (vial)	X	C	C
Metoprolol (injectable)	15 mg (vial)	X	X	R
Midazolam (injectable)	20 mg (prefilled syringe/vial)	X	R	R
Naloxone (injectable)	10 mg (prefilled syringe, vial, ampule)	R	R	R
Nicardipine (infusion)	N/A	X	X	TM

Medication and Formulation	Minimum Quantity	E	A/C	P
Nitroglycerin (sublingual)	4 mg (0.4 mg tabs, 0.4 mg metered dose spray, or lingual powder)	P	R	R
Nitroglycerin (infusion)	50 mg (premixed/vial)	X	X	R
Nitrous oxide (50/50)	N/A	O	O	O
Norepinephrine (infusion)	4 mg (vial/ampule)	X	X	R
Ondansetron (injectable)	8 mg (prefilled syringe/vial/ampule)	X	R	R
Oxymetazoline (nasal)	10 ml (bottle)	R	R	R
Pitocin (injectable)	Two (2) 20U (vial/ampule)	X	X	R
Phenobarbital (injectable)	780 mg	X	X	O
Phenylephrine (injectable)	1 mg/10ml (100 mcg/ml) prefilled syringe	X	X	R
Phenylephrine (infusion)	10 mg (vial/ ampule/premixed bag [10mg/250ml NS])	X	X	R
Phenylephrine (nasal spray)	15 ml (bottle)	X	X	R
Pralidoxime (autoinjector)	Service optional/if available	O	O	O
Pralidoxime (injectable)	Service optional/if available	X	X	O
Prednisolone (syrup)	120 ml bottle (5mg/ml)	X	X	R
Prednisone (tablets)	Three (3) 20 mg tablets	X	X	R
Procainamide (injectable)	Two (2) 1 gm (vial/ampule)	X	X	R
Promethazine (injectable)	12.5 mg (vial/ampule)	X	X	R
Proparacaine 0.5% (ophthalmic)	30 ml (bottle)	X	X	C/R
Pseudoephedrine (oral)	Service optional/if available	O	O	O
Rocuronium (injectable)	200 mg (vial/ampule)	X	X	R/C
Saline 0.9% (injectable)	One (1) 1000 ml bag	X	R	R
Saline 0.9% (injectable - for admixtures)	Two (2) 500 ml, Two (2) 250 ml, Two (2) 100 ml, Two (2) 50 ml bags	X	X	R
Saline 3% (injectable)	500 ml (bag)	X	X	R
Sodium bicarbonate (injectable)	150 mEq (prefilled syringe)	X	R	R
Sodium thiosulfate (injectable)	Service optional/if available	X	X	O
Succinylcholine	200 mg (any combination of vials)	X	X	O
Terbutaline (injectable)	1 mg (vial/ampule)	X	R	R
Tetracaine 0.5% (ophthalmic)	2 ml (bottle)	X	X	R/C
Thiamine (injectable)	100 mg (vial/ampule)	X	R	R
Tissue Plasminogen Activator (tPA)	N/A	X	X	TM
Tranexamic acid (injectable)	1 gm/10 ml (vial/ampule)	X	X	R
Vecuronium (injectable)	10 mg (vial - requires reconstitution)	X	X	R/C

Legend

X -   Not within scope
R -   Required
C -   Choice (interchangeable with another agent; service selects agent of choice)
O -   Optional (within scope if available)
P -   Within scope from patient's prescriptive stock
TM -   Within scope to monitor and titrate a pre-existing infusion

Interchangeable Medications

Methylprednisolone (injectable)	Hydrocortisone (injectable)	Dexamethasone (injectable) [required for P level]
Lorazepam (injectable)	Diazepam (injectable)	
Calcium chloride 10% (injectable)	Calcium gluconate 10% (injectable)	
Tetracaine 0.5% (ophthalmic)	Proparacaine 0.5% (opthalmic)	
DuoDote Antidote Kit	Mark1 NAAK	
Haloperidol (injectable)	Droperidol (injectable)	
Rocuronium	Vecuronium	

*Labetalol is required for P level services transferring patients receiving t-PA for Acute Ischemic Stroke.

Standard Concentrations for IV Admixtures

Medication	Admixture	Yield	Level	Pump
Amiodarone (bolus)	150mg/100ml D5W 300mg/100ml D5W	N/A	AC/P	NO
Amiodarone (infusion)	360mg/200ml D5W (premixed)	1.8 mg/ml	P	YES
Cefazolin	≤ 1 gm/50ml D5W, NS ≥ 1 gm/100ml D5W, NS	20 mg/ml	P	NO
Dextrose 10%	Premixed 10% solution	10 gm/100ml	AC/P	NO
Diltiazem	100 mg/100 ml D5W, NS	1 mg/ml	P	YES
Dopamine HCL	400mg/250ml (premixed) 800mg/500ml (premixed)	1600 mcg/ml	P	YES
Epinephrine	1mg/1000ml NS, D5W	1 mcg/ml	P	YES
Hydroxocobalamin	5gm/200m NS	25 mg/ml	AC/P	NO
Intralipid 20%	Premixed 20% solution	N/A	P	NO
Labetalol	N/A	N/A	P	YES
Lidocaine	1gm/250ml D5W (premixed) 2gm/500ml D5W (premixed)	4 mg/ml	P	YES
Magnesium Sulfate	2gm/50ml D5W, NS 4gm/100 ml D5W, NS	40 mg/ml	P	NO
Nicardipine	N/A	N/A	P	YES
Nitroglycerin	50mg/250 ml (premixed)	200 mcg/ml	P	YES
Norepinephrine	4mg/250ml D5W 8mg/500ml D5W	16 mcg/ml	P	YES
Pitocin	20U/1000 LR, NS	20 mU/ml	P	YES
Phenylephrine	20mg/250ml NS, D5W	80mcg/ml	P	YES
Procainamide	1.5gm/100ml NS, D5W	15mg/ml	P	YES
Saline 3 %	Premixed 3% solution	N/A	P	YES
Sodium bicarbonate	150 mEq/1000ml D5W (preferred) or 1000 NS	N/A	AC/P	No
Sodium thiosulfate	12.5gm/50ml	250mg/ml	P	NO
Tissue Plasminogen Activator (tPA)	N/A	N/A	P	YES
Tranexamic acid	1 gm/100 NS	10 mg/ml	P	NO

TABLE OF
CONTENTS